The Walter Lynwood Fleming Lectures in Southern History

LOUISIANA STATE UNIVERSITY

TOXIC DRIFT

PESTICIDES AND HEALTH IN THE POST–WORLD WAR II SOUTH

PETE DANIEL

Louisiana State University Press
Baton Rouge

in association with

Smithsonian National Museum of American History
Washington, D.C.

Manufactured in the United States of America
First printing

Designer: Laura Roubique Gleason
Typeface: Trump Mediaeval with Futura display
Typesetter: G&S Typesetters, Inc.
Printer and binder: Thomson-Shore, Inc.

Library of Congress Cataloging-in-Publication Data

Daniel, Pete.
Toxic drift : pesticides and health in the post–World War II South / Pete Daniel.
p. cm. — (The Walter Lynwood Fleming lectures in southern history)
Includes bibliographical references and index.
ISBN 0-8071-3098-2 (cloth : alk. paper)
1. Pesticides—Toxicology—Southern States—History—20th century. I. Title. II. Series.
RA1270.P4D36 2005
615.9'02'0975—dc22

2005002594

The paper in this book meets the guidelines
for permanence and durability
of the Committee on Production Guidelines
for Book Longevity of the
Council on Library Resources. ∞

To my granddaughters
Julie and Stella

CONTENTS

ILLUSTRATIONS

ACKNOWLEDGMENTS

Over fifteen years ago I worked on two exhibits at the National Museum of American History that featured pesticides. Researcher Louis Hutchins discovered a wealth of documentation in the National Archives and Records Administration that helped document not only a section of the Science in American Life exhibit but also an article on fire ants published in *Agricultural History.* A chapter in *Lost Revolutions* extended my examination of pesticides in the years after World War II. When asked by the Louisiana State University History Department to give the 2004 Walter Lynwood Fleming Lectures, I saw the opportunity further to expand investigations into pesticides and health.

Over the years I have accumulated many debts to people who have helped in this project. Some granted me interviews, others helped with research, and still others read and commented on drafts of the Fleming Lectures and/or the book manuscript. Archivist James Rush at the National Archives and Records Administration helped immensely, not only in locating records but also by introducing me to Edmund Russell, who became a colleague as a museum fellow. Archivist Joe Schwartz has expertly and patiently guided me through the gigantic U.S. Department of Agriculture, Agriculture Research Service, and Environmental Protection Agency collections in the National Archives. The staff of the Mississippi Department of Archives and History located the trial transcript for the *Lawler* v. *Skelton* trial, and the staff of the Southeast Regional Records Service in Atlanta located case files from several federal cases. Librarians at the National Museum of American History—in particular Chris Cottrill and Jim Roan—kept my project in mind and called my attention to books and articles, as well as tools such as ProQuest. Lawson Holladay and John McWilliams not only allowed me to use the plaintiff's records from the *Lawler* case but also talked to me extensively about the career of Pascol Townsend and their own experiences with pesticides.

Davina Chen, Smita Dutta, Steve Garabedian, Lawson Holladay, Jim Kelly, Carroll Leggett, John McWilliams, Deirdre Murphy, Katherine Ott, Edmund Russell, Alex Russo, and Jeffrey Stine read drafts of the Fleming Lectures. Monica Gisolfi's careful reading of the manuscript helped me reconfigure key concepts. Jessica Moore assisted with research and made helpful comments on the book manuscript. Mattie Sink and Martha Swain at Mississippi State University helped locate Dr. Mary Elizabeth Hogan for an interview. Marcel LaFollette generously shared with me her research at the Smithsonian Archives on the Science Service Records. Through an unlikely coincidence, Angela Less put me in contact with members of Charles Lawler's family. Emory University, where I was visiting professor in 2001–2002, gave me access both to the law library and to Lexis to search legal sources for pesticide cases. Norman Reigner of CropLife America discovered and shared pesticide-use statistics. I owe special thanks to Alisa Plant for skillfully copyediting the manuscript. My family has patiently humored me and tolerated my sometimes obsessive preoccupation with this project.

TOXIC DRIFT

1
THE REACH OF TOXIC DRIFT

These minute particles are the components of what we know as "drift"—the phenomenon that plagues every householder who receives contaminating spray from his neighbor across the street, or from his Government's spray planes several miles away. We are now beginning to wonder how vast the reach of "drift" may be.

Rachel Carson, June 4, 1963

Kids playing hide-and-seek in the DDT cloud trailing a spray truck is one of the most enduring, and haunting, images from the 1950s. Few people then gave much thought to the health implications of such games. Dr. A. E. Wood, a professor of chemistry at Mississippi College as well as the town of Clinton's mayor, did not know if DDT was harmful, but he believed "that any poison is dangerous." A truck sprayed DDT for mosquitos, and kids "would run along beside the stuff and inhale it." Dr. Wood assigned a marshal to keep kids out of the spray. While it might not cause immediate symptoms, he reasoned, "we don't know; it might be ten or fifteen years before it would show up." Dr. Wood offered this cautionary story as a witness in a 1960 trial involving human poisoning.[1]

The chemicals used on post–World War II farms and yards and in urban areas evolved from quickened war research on synthetic pesticides (insecticides, herbicides, fungicides, etc.), those created in the laboratory rather than found in nature. After being synthesized in the late 1930s by a Swiss chemist working for J. R. Geigy, DDT quickly emerged as a "miracle" insecticide that killed lice and mosquitos, the carriers of typhus and malaria respectively. DDT first arrived in the United States in August 1942; by the summer of 1945, production for military purposes reached 3 million pounds per month. Even during the war, chemical companies realized its vast postwar potential. In May 1944, D. F. Murphy, a research director for Rohm & Haas, invited the press to its experimental farm for a DDT demonstration. "You will hear how chemical science has marshaled its forces to help save the postwar world from much of the damage now done by insects," he boasted. The new family of synthetic chemicals, he explained, would "rid the world of mosquito-borne malaria," replace "such dangerous insecticides as lead and calcium arsenate," free the country from dependence on "imported pesticides," and help in "man's battle against insect invaders." Murphy presented a compelling picture of health, safety, and happiness. With the end of the war in August 1945, chemical companies such as Geigy, Du Pont, and Hercules aimed production at a domestic market eager to use DDT. Advertisers smoothly adapted war rhetoric calling for annihilation of foreign enemies to one that would exterminate insects and other pests.[2]

DDT became emblematic of synthetic insecticides. Created in the laboratory, heroically proven in war, introduced publicly with great fanfare, and deadly effective, it promised insect-free homes, lawns, gardens, and fields. Other synthetic insecticides, herbicides, and fungicides quickly lined store shelves. Corporate advertising effaced toxic warnings and featured smiling faces, dead insects, and astroturf-perfect lawns. Trusting both advertisements and federal oversight, few pesticide users paused to read labels carefully and thus remained unaware of a product's toxic core. Conventional wisdom held that pesticides were carefully tested and, while strong enough to kill insects, were weak enough to spare fish, wildlife, and humans. In fact, synthetic chemicals reached the market with hardly any testing for short-term (acute) or long-term (chronic) health effects. For a country consumed with fears of Soviet aggression, the Korean War, McCarthyism, polio, civil rights, Sputnik, and rock 'n' roll, pesticide dangers hardly registered—at least not until 1962, when Rachel Carson published *Silent Spring.*

This book deals primarily with three families of agricultural chemicals: chlorinated hydrocarbons, organophosphates, and herbicides. Chlorinated hydrocarbon insecticides—such as DDT, endrin, and heptachlor—seemed benign to humans, since they rarely caused acute health symptoms. Yet these chemicals retained their potency and spread through the environment with disturbing consequences. Organophosphate nerve poisons were designed by German scientists during World War II; after the war, corporations tailored formulations for the pesticide market. Organophosphates such as malathion and parathion were extremely toxic when applied but quickly lost their potency and became harmless within days or weeks. Even in the diluted concentrations used on insects, however, organophosphates sometimes sickened and occasionally killed people. Herbicides, or plant growth regulators (such as 2,4-D), which killed broad-leafed weeds and plants, were also developed by scientists during World War II. Herbicides quickly flooded the postwar market; advertisements touting them promised weed-free lawns and fields. Farmers immediately discovered that 2,4-D drift damaged such broad-leafed crops as cotton, and many complained of sickened cattle. The United States used Agent Orange (a combination of 2,4-D and 2,4,5-T) as a defoliant in Vietnam, with disastrous consequences both for the environment and human health. These three chemical families varied greatly in their toxicity and method of killing insects and weeds and in their threats to fish, wildlife, and people. Despite the lack of toxicity research, when chemical companies marketed chlorinated hydrocarbons, organophosphates, and organic herbicides, advertisements painted a world free of insects and weeds. The vision of a brave new chemical world often blinded both scientists and the public at large.[3]

In the quarter century after World War II, the commercialization of synthetic pesticides spread toxic residues across the globe, just as the nuclear destruction of Hiroshima and Nagasaki in 1945 and subsequent atmospheric tests both by the United States and the Soviet Union internationalized radioactive fallout. Riding air and ocean currents, invisible radiation and pesticide residues spread to the most isolated parts of the earth. By the middle of the twentieth century, every creature was living in the path of toxic drift. While critics of nuclear fallout and of pesticides warned of long-term health risks, defenders insisted that the dangers were overstated and that no proof existed of chronic health problems due to toxic drift. Indeed, because both radiation and chemicals lodge in the body and in some cases trigger disease years later, it has been difficult to trace illnesses to such exposure.

Long-term pesticide effects on fish and wildlife were not so vague as on humans. Chlorinated hydrocarbons not only killed insects on contact but also remained toxic; that is, they were persistent. Wind and ocean currents swirl toward the Arctic Circle, concentrating pesticide residues that are consumed by fish and wildlife. Paradoxically, some of the most desolate places on earth now have the most concentrated loads of toxic chemicals. Studies have revealed that Inuit women who live in arctic Greenland, Canada, Russia, and Alaska "have dangerously high levels of PCBs, DDT and mercury in their blood, fatty tissue and breast milk." Because they are at the top of the food chain, polar bears in Norway's remote Svalbard archipelago carry alarming amounts of toxic chemicals such as DDT and PCBs. Suspecting that chemicals are to blame for health problems, scientists are now studying bears' immune and reproductive systems.[4] After World War II, scientists failed to investigate the long-term possibilities of toxic drift but instead focused on inventions and new products. They were willing to allow the future to take care of itself.

Scientists emerged from World War II with enhanced prestige and power, and their research and development generated seductive products that advertisers easily transformed into dazzling symbols of a better life. Collateral costs seemed insignificant. In the scientific world, however, destruction often shadows creation, and in time pesticides produced toxic dumps, poisoned streams, wildlife kills, and human health concerns. But in the 1950s the day of reckoning seemed distant, and an inviolate notion of progress pervaded science and technology. This logic suggested that scientists could even solve science-generated problems, creating a closed loop that swallowed mistakes. If one scientist created a pesticide that had problems, another could rectify the flaw. Although scientists enthusiastically promoted this scenario, a consumer's everyday life became a treadmill, a perpetual patchwork of new products to replace old ones, one environmental and health threat followed by another.

Much like Don Quixote, with his well-intentioned but often disastrous chivalrous interventions, postwar scientists sometimes deluded themselves into thinking that they could save the world, whether it needed saving or not. Just as Don Quixote forced his heroic fantasies onto the most unlikely situations, hubristic scientists—in their rush to conquer nature—dismissed the implications of nuclear fallout and pesticide residues. But unlike the deluded Knight of the Sorrowful Face and his impoverished retainer, bureaucrats, scientists, and their savvy and powerful supporters enjoyed enormous financial support in their efforts to exterminate insects and kill weeds. The quest to destroy pests unfortunately outran research on insect resistance, ignored collateral environmental damage, minimized the dangers of residue accumulation, and downplayed threats to fish, wildlife, domestic animals, and, most importantly, to humans. Extolled by advertising and scientific enthusiasm, synthetic pesticides quickly became the ultimate weapon in the war against insects. Bewitched by their own ambition and rhetoric, some scientists hoped to reconfigure an imperfect world. In 1946, Roy Hansberry, head of Shell Agricultural Laboratory, typified scientific hubris when he declared, "We really doubt if there are really any healthy plants in nature." Nature was no longer good enough; nature was the enemy.[5]

While pesticide advocates pointed to the lack of data on health issues, critics argued that evidence showed dangerous trends that warranted caution. Much of the debate swirled around the Agricultural Research Service (ARS), the United States Department of Agriculture (USDA) unit that functioned as the clearinghouse for pesticide approval, as a scientific center that investigated possible problems, and as the sponsor of large insect-control projects. For twenty-five years, until the establishment of the Environmental Protection Agency (EPA) in 1970, the ARS administered U.S. pesticide policy. Within the ARS, the Pesticide Regulation Division (PRD) registered formulations, approved labels, tested products, and policed the marketplace. Seduced by the notion that farming should be primarily a business and not a way of life, USDA bureaucrats dreamed of capital-intensive farm units that utilized the latest science and technology. Like demented physicians, ARS administrators prescribed chemicals indiscriminately, usually without full knowledge of the target pest and seldom with any thought of side effects. Significantly, the ARS's eradication schemes, oversight, and labeling reflected the chemical industry's agenda. The extent of government-industry collusion only emerged publicly in the late 1960s, just before pesticide regulation was relocated to the EPA. Still, the ARS shaped the way pesticides were approved, labeled, and used in the quarter century after World War II. Its stewardship came under terrific criticism, but this rogue bureaucracy continued to champion chemical solutions long after serious environmental and human dangers emerged.

It was not inevitable that chemical solutions would win total victory in insect and weed control. As scientists were developing plant regulators during World War II, experiment stations in Mississippi and Alabama refined a tractor-mounted flame cultivator that destroyed weeds. Like a blowtorch, the flame singed young weeds but spared the tougher cotton stalks. Chopping out weeds was a major labor-intensive task in cotton cultivation, as was hand picking. Mechanical and/or chemical solutions to these labor needs had the potential to revolutionize cotton production. During the war, International Harvester developed a successful mechanical cotton harvester at the Hopson Plantation outside Clarksdale, Mississippi, and within fifteen years the machine had largely replaced hand pickers. For a time it seemed that flame cultivation might compete with chemical herbicides to clear weeds. In August 1947, the Science Service's Frank Thone, who had been told that 2,4-D drift had resulted in "a handsome crop of lawsuits," asked Flame Cultivation Corporation's Atherton Richards about herbicides. Having just returned from the Mississippi Delta, Richards reported that researchers at the Delta Experiment Station had applied one ounce of 2,4-D to a twenty-foot-square cotton plot, which "served to shrivel up half a field of cotton a hundred yards away." He reported that "merely the storage of 2,4-D in a warehouse resulted in a circle of damage extending 5 hundred feet around the warehouse." Richards, of course, was not a disinterested observer. Ultimately herbicides won out, despite the lack of research on health and the calamitous collateral damage from drift. The Reverend John Harris summed up the legacy of herbicides to oral historian Lu Ann Jones in 1988, pointing out that herbicides kept the sugarcane fields near Franklin, Louisiana, weed free. As he understood it, people at Louisiana State University developed the herbicides. "That's how they got rid of all the hoes," he said. "You don't see a hoe now unless you see it around somebody's house. They don't hoe no more."[6]

As USDA scientists promoted pesticides, their bureaucratic domain expanded to testing and regulating new products. The rapid expansion of

Advertisement for Black Leaf in *Progressive Farmer,* April 1, 1953, 80.

pesticide use on crops and yards and by ARS control programs swelled bureaucratic power. The first postwar USDA *Yearbook of Agriculture,* distributed in 1947, featured articles that recommended synthetic chemicals for fields, yards, and homes. The 1952 *Yearbook* featured insects, warning of the dire threat they posed to agriculture. In their eagerness to legitimate synthetic chemicals, USDA bureaucrats embraced ill-considered eradication projects and denied their unhealthy consequences. Single-minded, ambitious, and eager to curry favor with anyone higher in the USDA organizational chart, in Congress, or among chemical company executives, ARS leaders shamelessly and sometimes unethically promoted pesticides. Earlier scientific research on biological control fell away. With the fervor of the newly converted, ARS scientists and administrators worshiped pesticides as a plague-ending god. Bombarded by advertising and pressured by extension agents and salesmen, farmers eagerly stepped onto the chemical treadmill that had been set in motion by the development of synthetic chemicals.[7]

In a larger sense, the chemical revolution merged with mechanization and, after 1933, increasing government intrusion into agriculture. Planners within the USDA seized on science and technology, advocating capital intensive farm operations that drove off millions of farm laborers. In addition, the problem of the color line, as W. E. B. Du Bois had predicted at the turn of the twentieth century, was evident at midcentury; the USDA singled out black farmers active in civil rights activities and denied them loans and other benefits. The millions of displaced farmers represented the fruits of USDA policy to streamline farm operations. Hand laborers were embarrassing reminders of the inefficient past. As synthetic pesticides became an integral part of the agribusiness blueprint, pesticide use increased exponentially. Pesticides invaded not only the earth, air, and water, but also animal and human bodies.

Pesticides had been part of a farmer's arsenal long before World War II. For centuries farmers had studied ways to discourage insects from destroying their crops, and by the twentieth century they used Paris green, lead arsenate, nicotine, pyrethrum, and a variety of other compounds. With the increasing use of pesticides and fertilizers, politicians called for state and federal regulation both to protect the consumer and to prevent fraud. Upton Sinclair's exposé of the meatpacking industry, *The Jungle,* prompted the Pure Food and Drugs Act in 1906; four years later the Federal Insecticide Act regulated fraudulent practices. The 1938 Pure Food, Drug, and Cosmetic Act required manufacturers to submit test results for new products and, for the first time, allowed the Food and Drug Administration (FDA) to set legal tolerances, that is, the amount of pesticide residue allowed on food. The Public Health Service provided research on residues. Despite the laws and bureaucratic machinery, pesticide regulation was not an exact science. In essence, the system attempted to safeguard the public from dangerous pesticide residues in food. By controlling appropriations, congressional committees played a major role in shaping and enforcing pesticide laws. Mississippi congressman Jamie Whitten championed pesticides and, as chair of the subcommittee on agricultural appropriations, bent pliable USDA bureaucrats to his will.[8]

While pesticides used before the war could be lethal if taken in concentrated form, the dust formulations applied at ground level or from aircraft apparently caused few health problems; nor did they harm adjacent crops. The proliferation of synthetic chemicals tested the regulatory structure. The Federal Insecticide, Fungicide, and Rodenticide Act of 1947 required manufacturers to vouch for a product's safety and effectiveness, but if the USDA rejected a registration request, companies could register products under protest. To remove the product from the market, the USDA had to take the manufacturer to court and prove the product's danger. This loophole ensured that nearly all formulations submitted to the ARS would reach the market. In 1950 Congressman James V. Delaney held extensive hearings on pesticide safety, and chronic dangers from pesticides became a more pressing issue. In 1954 the popularly labeled Miller Amendment, offered by Congressman A. L. Miller, empowered the FDA to set tolerances at a safe level for specific uses.[9]

Despite synthetic chemicals' heightened toxicity, their defenders insisted that without pesticides U.S. agriculture would shrivel, insects would triumph, and the world food supply would be jeopardized. For the most part, the implications of toxic drift were ignored. Even as Congress and the federal bureaucracy attempted to control agricultural chemicals, pesticide advocates reinvented insects as voracious, pernicious, and threatening enemies that deserved extermination. Wheeler McMillen's 1965 book *Bugs or People?* portrayed insects as ghastly creatures worthy of a horror film: "Chewing and biting; crunching and sucking; puncturing and cutting; tunneling and boring, flying and buzzing; crawling and creeping; squirming and wiggling; dirtying and defiling; stinging and blistering; sapping and debilitating; corrupting and crippling; fouling and infecting; wounding and killing; tainting and contaminating; webbing and stunting; forever breeding, hatching, reproducing, multiplying; underground, above ground and in the air; attacking roots, stems, leaves, bark, wood, blossoms, grain and fruit; hiding, often unseen or invisible." After this neo-Faulknerian description, McMillen warned, "No farm is free from them. No acre, no square foot. No crop is exempt; no grain, no grass, no vegetable, no tree, no animal." McMillen presented this chilling insect profile as if farmers had not coexisted with insects for centuries.[10]

Still, public concern about chemical dangers to domestic animals and wildlife prompted both tests and regulation. By 1957 there were over two hundred basic chemicals in some eighty thousand registered formulations on the market. Guidelines setting residue tolerances in meat, milk, and other food provided wishful assurances that, if used properly, synthetic chemicals offered no threat to humans or wildlife. Statistical poisoning data was "scanty and misleading," according to Bernard E. Conley's 1957 *Journal of the American Medical Association* (*JAMA*) article, "and this has contributed to the misconception that pesticide injuries are infrequent or rare." Government residue research and guidelines dealt with the acute effects of such pesticides. The short-term consequences of chlorinated hydrocarbons such as DDT seemed benign, at least to humans, but long-term impacts only emerged in time. Indeed, scientists are still debating the question.[11]

The ARS generally ignored toxicity questions. Its projects to control or eradicate gypsy moths, bark beetles, and fire ants often relied upon large aircraft that sprayed millions of acres. During the fire ant eradication campaign in 1960, ARS scientists pondered why, in aerial applications of heptachlor and chlordane, a swath (one pass by the plane) had "excessive variations" of "tenfold or more." Some parts of a field had "excessive amounts and others practically none." When insufficient granules reached the ground, the ants were not bothered, much less eradicated. "While much work has been done to obtain uniform application across the swath width of planes," the report concluded, "apparently little is known about variation in a line parallel to the line of flight." With such shoddy research, eradication was at best a dream and at worst a hoax.[12]

Because drift was unavoidable, a body of common law emerged establishing cause of action and responsibility. Farmers who claimed crop or livestock loss could seek damages from landlords or pesticide applicators. Most states licensed pilots and required applicator companies to carry insurance and undergo inspections. Such statutes sought to balance the rights of aerial applicators, farmers who contracted for spraying, and neighbors.[13]

As lawyers battled liability issues, physicians followed cases of pesticide poisoning in *JAMA*. When DDT and other chlorinated hydrocarbons first came on the market in 1945, there were no studies of either their short- or long-term health effects. Since it was a poison, obviously DDT in large doses could cause sickness or death. The journal reported a case of twenty-eight "Formosan prisoners of war," who, when denied an evening meal, stole flour laced with 10 percent DDT and made biscuits. "Vomiting, numbness and partial paralysis of the extremities, mild convulsions, loss of proprioception and vibratory sensation in the extremities, and a hyperactive knee-jerk reflex were immediate toxic effects noted, but all these effects were transient except in 3 patients who had eaten excessive amounts," the article cryptically reported. Naturally there were concerns about people who worked in the formulating plants. A twenty-four-year-old worker making DDT aerosol bombs disregarded safety precautions, and "for four years he was exposed to vaporizations of the DDT formula for eight hours every working day." In May 1950 he consulted a physician, complaining of "weakness of four years duration, poor appetite and restless sleeping." Physicians discovered "severe fatigue, difficulty in talking clearly, glare phenomena, photophobia, blurring of vision and a feeling as though he were swimming and walking on air." Treatment and rest restored his health, and he returned to work. DDT could also prove fatal. On August 1, 1946, a fifty-eight-year-old laborer in good health accidently drank 120 cc of a 5 percent solution of DDT. Within an hour he began suffering "severe epigastric pain, and he vomited copious amounts of 'coffee ground-like material.'" These symptoms continued, and on August 6 he became comatose. He died a day later. *JAMA* also carried reports on organophosphate poisoning and a number of articles relating to antidotes.[14]

Given the number of poisoning cases, pesticide users must have been either extremely careless or uninformed about toxicity. Between March and October 1951, for example, Florida reported twenty-eight suspected parathion poisoning cases. By inhibiting cholinesterase activity, parathion and other organophosphates basically shut down the nervous system. Physicians recommended regular cholinesterase tests for people working around organophosphates. Two Florida physicians treated a nineteen-year-old parathion mixer who each day wore "clean coveralls, natural rubber gloves and a hat, and a respirator" and showered at the end of the day. On

July 20, 1952, he became ill. The employer stated that "poisoning was the result of accidental or careless ingestion of a relatively large amount of the chemical." The factory superintendent "found the patient somewhat stuporous, perspiring freely, and complaining of nausea, vertigo, diminution of vision, and headache." Shortly after being hospitalized, he became comatose and was treated with the antidote atropine. Slowly the patient improved, regaining consciousness after four days of treatment. "Although the cholinesterase levels returned to normal," the physicians reported, "it was obvious that the patient was confused, somewhat amnesic, and definitely child-like in his reactions and attitudes." He remained hospitalized until August 7.[15]

In 1954, the American Medical Association's (AMA's) committee on pesticides reported on household chemicals, stressing that many such chemicals were especially dangerous to children between one and three years old. The committee found a "lack of adequate labeling of many products freely used in the home," an issue that would persist for years. "Many of these products contained agents that were recognized as offering industrial and occupational hazards," the report continued, "but were now being used by persons unaware of their potential harmfulness."[16] Despite numerous articles chronicling pesticide poisoning and a report based on 37,000 "incidents of poisoning reported by poison control centers, industrial commissions, and state departments of health for the period covering 1956–57," the AMA's committee on pesticides came down surprisingly softly on ARS control programs. In 1958, the AMA committee observed that residents in New York, New Jersey, and Pennsylvania sought an injunction to prevent indiscriminate spraying of 3 million pounds of DDT. "The fire ant control program presents a similar problem in the South," the committee observed. Still, the committee reported, "It was concluded that opposition to spraying programs is often caused by persons who do not know enough about such programs and are frightened by what they do not understand."[17] The fact that residents sought an injunction to halt the indiscriminate spraying of 3 million tons of chemicals on populated areas suggests that people did understand the issue.

Although the ARS and its pesticide appendages successfully crushed opposition to their control programs, files in the National Archives and Records Administration contain thousands of letters relating to wildlife and human health. In the years before the publication of Carson's *Silent Spring,* complaints chronicled poisonings, called for investigations, and raised significant health questions. The USDA invariably defended pesticides as benign. The Reverend Gilbert P. Herrman complained from Dodge City, Kansas, to President Dwight D. Eisenhower about mass spraying and other kinds of pollution, and he received a typical ARS response. W. L. Popham, a staunch ARS pesticide defender, prepared a fact sheet for Byron T. Shaw's response praising DDT and other chemicals that had increased agricultural productivity. "Insecticides such as DDT, chlordane, and lindane which are employed extensively have meant much to citizens in our country and to people throughout the world," Popham boasted. "Millions of lives have been saved or illnesses prevented by destroying insects which transmit diseases of man." Herrman dismissed Shaw's response. "Everybody knows that the pesticide manufacturers are in a multi-million dollar business and that they will pull every string possible even at the expense of human life in order to get that 'almighty dollar,'" he replied in July 1953. "It makes us wonder which is the more important dead bugs or living human beings." This letter elicited a three-page reply from ARS acting administrator, M. R. Clarkson. He assured Herrman that lindane vaporizers and chlordane had been thoroughly studied and deemed safe if used as directed. Pesticide manufacturers, Clarkson wrote, eagerly cooperated with the ARS "in determining safe use for the products they intend to market."[18] This exchange, one of thousands in the quarter century after World War II, epitomized bureaucratic justifications. Significantly, by 1970, DDT, chlordane, and lindane had been outed as serious health threats both to wildlife and to humans. Marketed as miracle chemicals without research on their chronic physical effects, these and many other formulations remained on the market even after research raised serious questions about them.

In *Lost Revolutions: The South in the 1950s,* I discussed the USDA's failure to make public the illegal chemical residues in milk and beef and thus baby food. By embracing and defending chemicals, ARS leaders not only abandoned health concerns but also entered the domain of bureaucratic dereliction. ARS bureaucrats feared public reaction if residue studies were made public. The report that documented residues in milk was ultimately "sent to files." T. C. Byerly, ARS assistant director of livestock research, suspected that the report, if released, could lead to "a great deal of public clamor," adding, "Such attention has been directed with respect to DDT, with respect to antibiotics and with respect to fall-out." The USDA cloaked health concerns by admitting nothing and by promising new studies. The unfinished study became a staple not only of ARS policy but also a chemical company weapon, a strategy used to delay decisions while promising scientific objectivity.[19] The ARS protected agribusiness units (such as beef and dairying) from unfavorable publicity by muting public health issues. Ironically, physicians, scientists, and bureaucrats within agribusiness knew far more about toxicity than those outside the industry, but they hoarded or slanted their information—with dire consequences.

After World War II, synthetic chemicals emerged as a crucial component in the evolution of corporate farming. In an unintended double entendre, a Science Service article published just after World War II proclaimed, "Only a very poor farmer does not use a variety of chemicals to help him raise his crops these days." Unlike mechanization and government subsidies, which simply forced small farmers off farms, synthetic pesticides both reduced the demand for labor and threatened the health of those left on the land. The agribusiness infrastructure stretched from county USDA offices to state agriculture departments, from chemical companies to the ARS, from federal experiment stations to land grant universities, and from implement companies to such lobbying powers as the Cotton Council, the National Agricultural Chemicals Association, the Delta Council, and the American Farm Bureau Federation. When united, these components possessed enormous financial and political power. The agribusiness agenda ignored collateral damage to wildlife and downplayed human health concerns. Acting in concert, agribusiness interests quashed residue reports, rushed in experts when chemicals were suspected of causing health problems, and ensured that the emerging capital-intensive structure succeeded.[20]

2
POISONING

A complete clinical study of this patient should have produced a satisfactory diagnosis other than alleged insecticide poisoning.

Dr. Griffith E. Quinby

From his description of the episode it is rather amazing in fact that he lived through it.

Dr. Watts R. Webb

On the morning of August 16, 1956, the people around the Marie Gin and the surrounding cotton fields in Sunflower County, Mississippi, were routinely going about their work. Gin manager Charles Lawler and his African American helper, Johnnie B. McCaleb, were welding steel supports on a concrete slab beside the gin house. Lawler had managed the Marie Gin for six years and spent the off-season building and repairing gins. Thirty-four-year-old World War II veteran Tracey Skelton had moved to Marie in 1951, renting five hundred acres from V. A. Johnson, who also owned the gin. Landlord and tenant had walked the cotton fields the day before and had decided to spray for boll weevils. About nine o'clock in the morning, Skelton stopped by the gin and told Lawler that one of J. L. Turk's crop dusters would be spraying later that morning. Lawler insisted that the spraying be delayed until he completed the welding job. Like many people who lived in rural America, he thought it prudent to avoid chemical spray. Later Lawler heard the Turk Dusting Service Stearman in the distance and assumed that pilot John Martin had been told to avoid the gin area. Skelton, however, had never gotten around to warning Martin.

On the last of his three flights that morning, John Martin began the tricky approach to the edge of the cotton field by the Marie Gin. He drifted under power lines, narrowly missed the gin, then increased power and turned on the spray. Below Martin's plane, fifty-year-old Charles Lawler knelt on the concrete slab, intent on his welding. Johnnie McCaleb saw the plane at the last moment and shouted a warning to Lawler before stepping

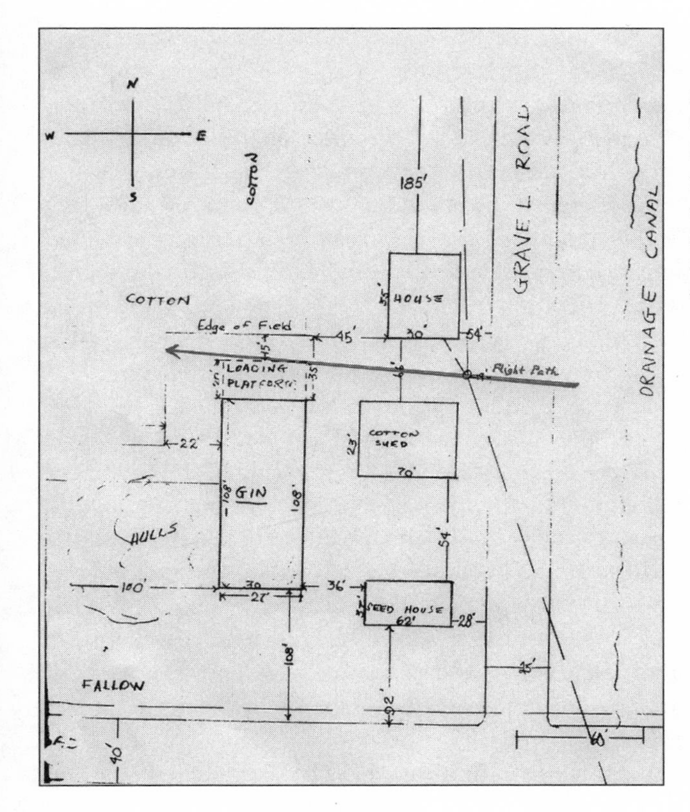

Map of Marie Gin by Charles S. Lawler, March 5, 1959.

Lawler *v.* Skelton *trial transcript (no.7868), 241 Miss. 274 (1960), Mississippi Department of Archives and History.*

under the gin shed. Before Lawler could react, a cloud of garlic-smelling insecticide swept over him. He began choking and coughing and tearing at his welder's helmet; he could not get his breath. McCaleb helped him get the welding hood off and then swabbed the poison from his face with cotton. Lawler went to his nearby house for lunch, where he drank some milk. He complained of nausea and the persistent taste of pesticide, but he stubbornly returned to the gin. He spent a wretched night twitching, fighting a high fever, nauseated, and vomiting. The next morning he attempted to work but soon returned home. He lay down to rest and went into a coma.[1]

Dr. A. A. Aden arrived at the Lawler home a short time later, and Lawler's wife, Irene, explained that her husband had been poisoned by the chemicals sprayed the day before. Dr. Aden immediately sent Lawler to the Indianola hospital and treated him there. After regaining consciousness, Lawler was nervous and dizzy; he had difficulty breathing and ran a temperature as high as 104.5 degrees. Blisters appeared on his shoulders and neck. He continued to taste the poison. On August 20, Lawler returned home and attempted to work at the gin, but weakness, nausea, and nervousness incapacitated him. As the ginning season progressed, Irene Lawler, with guidance from her husband, kept the books and managed the Marie Gin. Charles Lawler lingered at the gin as long as his strength permitted, but continued weakness, a low-grade fever, coughed-up blood, and nausea forced him home to bed. "I couldn't do manual work," he complained. He took medication that eased his symptoms and allowed him to sleep. His eyes would not focus, so he went to an eye specialist who "fitted me with glasses and had to change them three or four times before it got to where I could see." Prior to the poisoning, Lawler had climbed in his gin work and in construction. After the poisoning, he complained, "I was dizzy." He could not stand heights. "Just stepping up off of the floor seemed like I was 100 feet from the ground so far as the effects of being elevated," he remembered. His throat hurt and his low fever continued. He recalled that a doctor told him that "my throat looked like a piece of beef steak." When Tracey Skelton visited Lawler in the hospital, he identified the chemicals as a mixture of malathion, endrin, and the solvent xylene. He also told Lawler that his insurance did not cover these chemicals, suggesting that there would be no financial help with hospital bills.[2]

Malathion, an organophosphate, and endrin, a chlorinated hydrocarbon, were two of the numerous synthetic chemicals developed during World War II. Endrin became a popular pesticide, although scientists traced it to fish kills in the Mississippi River in 1963–64 and later associated it with human health problems. Endrin poisoning symptoms include headache, dizziness, nervousness, confusion, nausea, vomiting, and convulsions. It was banned in the United States in 1986. Despite being highly toxic and causing numerous accidental deaths, malathion and parathion enjoyed widespread use as pesticides in the postwar era. Malathion poisoning symptoms include "abdominal cramps, vomiting, diarrhea, pinpoint pupils and blurred vision, excessive sweating, salivation and lacrimation, wheezing, excessive tracheobronchial secretions, agitation, seizures, bradycardia or tachycardia, muscle twitching and weakness, and urinary and fecal incontinence." Since neither chlorinated hydrocarbons nor organophosphates dissolved readily in water, they were mixed with a solvent. Xylene was commonly used as a solvent, paint thinner, and gasoline additive; it could

Charles Lawler with daughters Jean (right) and Sue (left).

Compliments of Sue Culver.

cause "headaches, lack of muscle coordination, dizziness, confusion, and changes in one's sense of balance."[3] The petroleum solvent in the malathion mixture that affected Charles Lawler was not identified. Scientists had also discovered potentiating reactions when several chemicals acted in concert to produce an effect greater than the sum of their parts. No tests were done on the mixture that was accidentally sprayed on Lawler.

With the help of his wife, Lawler got through the ginning season. On December 13, he drove to Greenville to renew his car license plates. "I got out of the car and straightened up," he recalled, "and it was just like something snapped—there was heavy pressure in the right side. I must have blacked out a minute or two." Not knowing any doctors in Greenville and fearing a long wait at the Gamble Brothers Clinic there, Lawler drove back to Dr. Aden's office in Indianola, only to discover that Dr. Aden was playing golf. The nurse assisted him and sent him to the South Sunflower Hospital, where he remained for four days. Although Dr. Aden suspected a heart problem, he admitted to Irene Lawler that he had no satisfactory explanation of her husband's continuing illness. Lawler's health did not improve, and in late December Dr. Aden sent him to King's Daughters Hospital in Greenville for a checkup. Dr. B. F. Hand found no heart problem and diagnosed Lawler's immediate problem as a collapsed right lung. On January 2, 1957, Lawler returned home with instructions to remain in bed for seventeen days, taking penicillin and sedatives. Ten days later a grease fire on the cook stove smoked the house and caused him to have a coughing spell. Irene Lawler rushed her husband to the hospital, "as it seemed he first stopped breathing." He spent sixteen days in South Sunflower Hospital. "I was choking and smothering most of the time," he recalled. He bought an oxygen mask and tank, which he kept by his bed. In June he went to University Hospital in Jackson for the full checkup required by U.S. Fidelity & Guaranty insurance company for workman's compensation applicants.[4]

In August 1957, Lawler sought treatment from Dr. Mary Elizabeth Hogan in Glen Allan. Born in Starkville, Mississippi, in 1916, Hogan was a medical technician for six years before realizing her ambition to become a physician. She attended the University of Mississippi Medical School for two years and then completed her training at the University of Tennessee Medical School. In 1952 she established a clinic in Glen Allan, where she became widely recognized for her selfless practice among the black population around Lake Washington. She cared for patients, dispensed drugs, and was often paid, if at all, with chickens or eggs. Her one extravagance was a red Chevrolet convertible that, incidentally, allowed patients instantly to locate her in emergencies. In the summer and fall of 1957, Dr. Hogan treated many workers in the area for parathion poisoning, for the air was thick with residue. Her friend Helen Neal recalled that "late in the afternoons after they had poisoned it was just a haze over the lake." Later the lake was closed to fishing because of pesticide buildup. Evidently Charles Lawler consulted Hogan because he had heard of her work in Glen Allan. His lingering sickness, Hogan diagnosed, was the result of poisoning. The number of sick black workers affected by parathion, she revealed, "was distressing." While she might normally see 25 or 30 African American patients a day, when the poisoning season began in 1957 that number increased to 125 a day. Her patients arrived with high fevers. Confused, coughing, and spitting up blood, they would lay out in the yard until she could see them. She diagnosed their malady as chemical pneumonia, treated them with atropine, and advised them to stay away from the fields during spraying.

Dr. Mary Elizabeth Hogan.

Compliments of Dr. Hogan.

Dr. Hogan understood that organophosphates blocked nerve impulses, and she reported that people had died of respiratory failure as a result.[5]

The press seized on reports of worker poisoning. Several stories on the subject appeared in the *Greenville (Miss.) Delta Democrat Times,* and radio advisories publicized it as well. In mid-August, Dr. Hogan advised tenants whose homes were surrounded by cotton fields to take their children and leave the area during poisoning. The Public Health Service's Griffith Quinby, who was researching pesticide toxicity in the delta, diagnosed only one case of organophosphate poisoning. Dr. R. W. Williams, the county health officer, urged farmers to "proceed with normal precautions." A week later state health workers took blood samples, presumably to check cholinesterase levels, although such tests were unreliable unless conducted immediately after exposure. In the meantime, eight hundred African American students in Carroll County were sickened with what health authorities labeled Asian flu.[6]

On September 10, *Business Week* correspondent John Donaldson contacted M. R. Clarkson at the Agricultural Research Service (ARS), asking about a report that sixty persons near Glen Allan were stricken with fevers up to 105 degrees and pneumonia-like symptoms. The local physician, Donaldson continued, had diagnosed poisoning from parathion and xylene. Clarkson arranged for an investigation that afternoon by a Public Health Service (PHS) representative, who with indecent haste concluded the same day "that the condition was in no way related to the use of insecticides." The representative forwarded the report to the PHS's Wayland J. Hayes Jr., and Clarkson informed Donaldson to contact Hayes if he needed more information. At this time, Hayes was emerging as a supreme apologist for the chemical industry, deflecting numerous inquiries about pesticides and health. In time, the Glen Allan poisonings were conflated with the reports of Asian flu. Six years later, for example, Dr. A. L. Gray, executive director of the Mississippi State Board of Health, stated that the Glen Allan poisonings were "one of the first outbreaks of Asian flu in Mississippi." There were numerous Asian flu cases in the state that year, but Dr. Hogan was positive that her patients had been poisoned, for they responded positively to atropine treatment. When the PHS representative concluded that they suffered from Asian flu, Dr. Hogan recalled, "I just felt like I was lost because he was the man of the last word." The PHS often stepped in to minimize pesticide fears. A Sarasota, Florida, newspaper reported in October 1956 that malathion, sprayed to combat a Mediterranean fruit-fly infestation, affected people with asthma and caused skin rashes. The ARS immediately requested "a reassuring statement from the U.S. Public Health Service." Dr. Hogan was also disappointed that planters, many of them her friends, refused to take her seriously when she warned that parathion created major health problems. "Even though the people were falling out in the fields, the planters wouldn't let you believe that the parathion had anything to do with it," she remembered. Dr. Hogan's Glen Allan clinic was practically surrounded by cotton fields, and pesticides drifted through its windows. As spraying continued into autumn, she became nauseated and experienced "mental confusion and headaches and a feeling of being mixed up and not being able to say what I wanted to say." On October 5, 1957, she was forced to leave her practice and seek medical attention for a month at University Hospital in Jackson. After recuperating at her sister's home, she attempted to resume her practice in Glen Allan but got no further than Vicksburg before pesticide drift sickened her. She never practiced in the delta again. After studying psychiatry for two years at the University of Louisville, she headed the department of female services at Mississippi State Hospital at Whitfield until her retirement.[7]

Dr. Hogan's experience raises the likelihood that poisonings often went unreported. The planter elite did not want to call attention to health problems for fear of disrupting delta agriculture. Dr. Hogan's longtime friend Helen Neal, who raised mink in Glen Allan, recalled that every time there was spraying she lost a mink. She also observed that flagmen were constantly exposed to chemicals as they marked the fields for crop dusters. No one seemed to discuss the danger of pesticides, although she saw evidence of their effects. "But up in the Mississippi Delta, up around Glen Allan," she pointed out, "you can see the effects of all of that because there were so many widows around there." Planters would go out in the fields after poisoning, she added, "and about all the men passed away early." It is easier to understand why African Americans failed to complain than whites. As machines and chemicals increasingly replaced workers and as the Citizens' Councils stifled any protest, African Americans were loath to voice their concerns lest they lose their jobs. But whites were also endangered. Neal believed that the threat of economic ruin forced planters to risk their health and that of their workers. It was a way of life.[8]

There were other occasional glimpses into the problem. During poisoning season in September 1963, Mississippi speaker of the house Walter Sillers complained to the Mississippi State Board of Health's Dr. A. L. Gray.

Sillers received daily complaints about "the pollution of the air from large and continuous volumes of poisons put out by airplanes and other devices" to kill insects. "Many," he continued, "think these poisons are doing about as much harm to people as they are to the insects." Sillers, whose home was in Rosedale, complained that when he rode through the countryside during poisoning season, the chemicals affected his "eyes, nose, throat and ears." Although Sillers wanted to avoid bad publicity or alarm, he suggested that Dr. Gray investigate the problem. Delta farmers depended on pesticides, Sillers admitted, but he was concerned with human health. "I don't want to continue it if to do so will be dangerous and damaging to life and health of human beings—or to cattle, livestock and other domestic animals," he wrote.[9]

Dr. Gray replied that he had received numerous complaints about pesticides, especially from people in the delta. He mentioned an incomplete study done several years earlier, referring either to that of Griffith Quinby or Richard L. Fowler. "The subject is very complicated with many economic and possible health facets and I believe a long-term research project would be quite expensive," Gray continued, "but would be necessary to determine the possible long-term effect of air pollution from insecticides."[10] Sadly, despite the potential dangers, no objective state or federal study emerged. Even as Gray and Sillers exchanged letters, endrin from fields and factories was killing fish in the Mississippi and Atchalafaya rivers.

While the vast ecological changes that swept through the rural South were hardly noticed, increasing tension generated by the civil rights movement captured national attention. The punitive and racist Citizens' Councils began in Leflore County, Mississippi, and spread across the South. Whites used economic sanctions—such as cutting off bank credit and access to fuel, or manipulating U.S. Department of Agriculture (USDA) policies to deny black farmers loans—and more direct intimidation to exact vengeance on black farmers who attempted to register to vote or who belonged to the NAACP. Boswell Stevens, head of the Mississippi Farm Bureau Federation, openly supported the Citizens' Councils. In May 1955 the Reverend G. W. Lee was murdered in Belzoni, and six months later Gus Courts was wounded by a shotgun blast as he tended his store. Both men had been active in the civil rights movement. In August 1955 in Tallahatchie County, fourteen-year-old Emmett Till was murdered for allegedly whistling at a white woman. This violence occurred in counties bordering on Sunflower County. Throughout Mississippi, fear and dread possessed the white community. Mississippi's African Americans saw hope in the *Brown* v. *Board of Education* decision and the nascent civil rights movement; they also saw labor-intensive jobs eliminated by tractors, picking machines, and chemicals. In 1956, anxious blacks and whites in the Mississippi Delta confronted each other over issues they considered far more important than pesticide poisonings.

Despite his continuing weakness and other health problems, Charles Lawler attempted to work during 1957. He supervised repairs to the nearby Eastland Gin and oversaw repairs at the Marie Gin. The weekend before ginning season started, he recalled, V. A. Johnson "brought me a check and told me he would have a man in the gin house Monday morning." Lawler then worked in a supervisory capacity for the Adair Planting Company Gin. That was his last job. He consulted attorney Howard Davis for guidance with a workman's compensation filing. Lawler learned later that Davis failed to advise him that even if the poison had aggravated a preexisting condition, he could have collected $8,600 plus his hospital bills. Lawler settled for $1,200, nearly all of which went to pay medical expenses. Irene Lawler bitterly recalled that Johnson had opposed her husband's attempt to receive workman's compensation. After the 1957 ginning season, Lawler received $29 each month from the welfare department and $20 a month for each of his two minor children. In February 1958, he moved in with his adult son in Wilmot, where he spent most of his time in bed taking oxygen. During the summer and fall pesticide spraying season, he explained, "I have to keep my room closed and stay in the room right by my oxygen at that time all the time."[11]

Charles Lawler had not been well served by his lawyer or most of his doctors. In the summer of 1958, as his health continued to fail and his medical bills mounted, Lawler turned to Pascol Townsend, an attorney who practiced in nearby Drew. Townsend graduated from Ruleville High School, attended Sunflower Junior College for two years, and then entered the University of Mississippi Law School. World War II interrupted his education, and he became a naval aviator. After the war he completed his law degree and, after practicing briefly in Ruleville, joined William Taylor's firm in Drew. Townsend had a weakness for underdogs and cared more about practicing law than compensation. One of his younger partners, Lawson Holladay, went so far as to state that Townsend was primarily interested in helping poor people and only became interested in collecting fees when taxes were due in April. During the civil rights movement, Townsend

Pascol Townsend.
Townsend, McWilliams, and Holladay, Drew, Mississippi.

represented Mrs. Fannie Lou Hamer and her husband. "And they would tell you flat out that he was the only white lawyer they trusted," Holladay recalled, adding, "or the only lawyer they trusted." Townsend mentored both Holladay and John McWilliams, and they fondly remembered how he had rescued them from mistakes and had molded them as lawyers. Serious in the courtroom, Townsend was affable outside it, often inviting the opposing counsel for a drink or meal.[12]

Given his deteriorating health and increasing medical bills, Lawler decided to sue V. A. Johnson, Tracey Skelton, and J. L. Turk for damages. Elizabeth Hulen, an attorney with the Jackson law firm of Watkins and Eager, joined Townsend in the case. Hulen had earlier lived in the delta but had moved back to Jackson; before completing her law degree, she had worked as a secretary in the firm that her father, William Hamilton Watkins, had founded in 1895. It is possible that she knew Townsend from her days in the delta, and certainly Hulen was a family friend. Hulen was the first woman in Mississippi history to argue before the U.S. Supreme Court. Townsend and Hulen were a formidable team, and they began interviewing

Elizabeth Hulen.
Compliments of Watkins and Eager, Jackson, Mississippi.

the Lawlers, physicians, and witnesses. Hulen educated herself on medicine and pesticides in preparation for the trial. Forrest G. Cooper of Indianola, aided by Oscar B. Townsend and Roger Tuttle, headed the defense team. In notes written during the trial, Hulen complained, "Mr. Cooper—delayed this case nearly 2 years."[13]

At first Pascol Townsend and Elizabeth Hulen focused on the workman's compensation issue, for Lawler—who over his lifetime had managed a plantation, owned a welding and machine shop in Arkansas, and served as deputy sheriff, constable, and city marshal—insisted that he had been misled by attorney Howard Davis and "in reality forced to a settlement." As a skilled worker, he made about $600 a month, supplementing this salary by repairing and installing gins. When Townsend interviewed Lawler on August 9, 1958, his weight had dropped from 155 pounds to 112, and his present doctor "says that there is no question but that his present condition . . . might kill him." Lawler had cut back on his smoking, although he confessed that "when he gets nervous or sick he smokes more." By early September, the case had evolved into a suit for damages. When Hulen talked with Cooper about the suit, he seemed only mildly interested in the particulars but was curious about the role of V. A. Johnson. Hulen explained that the relation between Skelton and Johnson "was that of joint adventurers." She continued to accumulate statements from doctors, but on October 28, as she completed the case declaration, she confided to Townsend that Lawler "should be prepared for the fact that we do not anticipate as big a judgment as we had once hoped for." In February 1959, Townsend sent a copy of the brief to his friend Eddie Peacock for his reaction. Peacock confided that a friend "told me several weeks ago that the law suit is an unpopular one." Townsend disagreed, writing, "I think this is based on a misunderstanding of the law and circumstances of this particular case."[14]

While Townsend and Hulen prepared the plaintiff's case, Forrest Cooper constructed a defense for Johnson, Skelton, and Turk. Cooper was small and dapper, a sharp dresser and a notable lawyer. He was also well connected. Delta planters as well as the broader agribusiness community were concerned that a Lawler victory in court might set a precedent for pesticide suits, and they were determined to defend their interests. While Cooper's correspondence does not survive, there are clues as to how he put his case together. Shortly before the trial began in March 1960, Dr. E. F. Knipling, director of the entomology research division of the ARS, pondered a telegram from B. F. Smith, head of the Delta Council, which was a chamber of commerce–like body that supported planters in the Mississippi Delta. Charles Lawler, Smith reported, had brought a $150,000 lawsuit against the landlord, tenant, and aerial applicator for damages suffered "as the result of the application of a malathion-endrin mixture for the control of cotton insects." With the trial imminent, Smith suggested that Dr. Marvin Merkel, an entomologist at the Delta Branch Experiment Station in Stoneville, Mississippi, as well as a Shell consultant, testify as an expert witness. The ARS supervised the experiment station complex throughout the country. Merkel's testimony could be vital, Smith's telegram suggested, and the case "could have influence on cotton insect control program in future years." Another ARS staff member, Knipling wrote, had "pointed out the importance of this suit and the effect it could have on cotton insect and other insect control programs if the claim for damage is sustained, even though there is every indication that the insecticide was not responsible for the plaintiff's illness."[15]

Both B. F. Smith's request for expert testimony from Dr. Merkel (a government scientist with ties to Shell) and Knipling's solicitousness about the lawsuit epitomized the ties between government scientists and the private sector. As the use of synthetic chemicals increased, the close World War II relations between government entomologists and the military blossomed into a larger network. Smith also worked closely with Boswell Stevens, head of the Mississippi Farm Bureau Federation; both men were connected to Mississippi State University (the state's land grant school), to experiment stations, and to the network of county agricultural agents and USDA bureaucrats in every county in the state. As Mississippi cotton production evolved from labor-intensive plowing, chopping, and picking into capital-intensive tractors, pesticides, and picking machines, those who stood to benefit from the emerging system allied to promote their interests. Forrest Cooper had assembled a formidable list of expert witnesses, and no doubt the Delta Council, the Farm Bureau, the ARS, and chemical companies all cooperated to sanitize toxic fallout. The trial did not generate newspaper publicity, but word spread around Sunflower County that a verdict for Lawler would jeopardize delta agriculture. While the lack of publicity suggested a case lacking significance, the caliber of the legal counsel, the prominence of the expert witnesses, and the relevance of the issues situated it squarely in the national debate over pesticides and health. Community sentiment coalesced around agribusiness stakeholders and, with the unmatched bloodless efficiency of the rural and small-town South, public opinion turned against Charles Lawler.

Forrest Graham Cooper, according to longtime office assistant Charlotte Buchanan, was a colorful and intelligent lawyer. Cooper was born in Forest, Mississippi, and was educated at the University of Mississippi. Buchanan remembered Cooper as a slight, charming man, occasionally absentminded but spectacular in the courtroom. People would fill the Sunflower County courtroom to watch Cooper and other lawyers perform. Meticulous in preparation and a master at jury selection, Cooper was a formidable opponent. "Most black people and most people in the Delta called him a banty rooster, bantam rooster," Buchanan recalled. "And he was a feisty kind of little guy. And you either liked him or you hated him. There was no in between."[16]

The trial began on March 15, 1960, presided over by Judge John D. Greene Jr. from Starkville. Charles Lawler and his attorneys faced not only the defense attorneys but also, behind the scenes, the Delta Council, the Farm Bureau, the ARS, chemical companies, and other stakeholders. Only traces of this broader support emerged publicly, but judging by the caliber of his expert witnesses, Forrest Cooper was wired into powerful local and national interests that were accustomed to defending pesticides. Cooper set out to defend not only Johnson, Skelton, and Turk, but also agricultural chemicals. V. A. Johnson distanced himself from the case as much as possible, casting all responsibility on his tenant, Tracey Skelton.

Skelton was the first witness. Except for a stint in the army between 1942 and 1944, he had lived in the delta all of his life. In 1951, a year after Lawler became gin manager, Skelton moved to the Marie plantation and became Johnson's tenant. In 1956 he was farming five hundred acres with the help of seven tenants. Townsend and Hulen attempted to link V. A. Johnson to Skelton by charging that their landlord-tenant relationship was a joint adventure, but when asked about Johnson's supervision, Skelton declared, "I planted to suit myself." He admitted buying thirty-five gallons of malathion at $7.85 per gallon and forty-five gallons of endrin at $7.70 per gallon and delivering the chemicals to J. L. Turk's airfield, but he asserted that he did not supervise their mixing. Skelton recalled that he had talked with Lawler the morning of the incident but denied being asked to delay spraying near the gin until Lawler completed his welding.[17]

V. A. Johnson's Marie property was about seven miles northwest of Indianola. He testified that he had nothing to do with Skelton's day-to-day operation. Although witnesses placed him in the fields around the Marie Gin the day before spraying, he claimed that he had left for New Orleans with his wife on August 14 and produced gasoline receipts purporting to prove his claim. Johnson admitted to only the most vague association with Skelton.[18]

J. L. Turk moved from California in the early 1930s and began his spraying company in 1948. In 1956 he owned five Stearman biplanes. Turk's Stearmans were outfitted with 130-gallon tanks that could hold a thousand pounds of mixed chemicals; eighteen nozzles sprayed the chemicals. Turk kept no chemicals at his landing strip. Each customer delivered the prescribed chemicals to the landing strip and provided farm hands to mix the batches. When Pascol Townsend asked if it would be difficult to warn a pilot that something was wrong once he was in the air, Turk reasoned that a pilot would stop spraying if someone was waving in the field below. When asked if he would spray a field where workers were present, he replied, "We would have sprayed the field that they were in that particular year, because it wasn't recommended that they take them out." He had never heard of any sickness among flagmen who stood at the end of the cotton rows to guide pilots. Turk explained that the nozzles that sprayed the Skelton crop were "set for two gallons per acre" and that there was no cockpit adjustment to change it. Johnny Martin flew the plane, he continued, and Martin was a pilot with ten years' experience.[19]

Defense lawyer Oscar Townsend asked Turk to examine a bill dated August 8 and August 11, 1956, for two pesticide applications on the Skelton crop. Turk admitted that he had changed the second date from August 10 to August 11. "Can there be any possibility it was the 16th?" Forrest Cooper later asked. "No," Turk replied. Had the spraying occurred on August 11, this line of reasoning suggested, then Turk was not liable.[20]

John Martin, who flew the Stearman on August 16, had been spraying chemicals since he was sixteen years old. He had flown for National Airlines for nine years, and at the time of the trial he was piloting Lockheed Electra turbojets out of Miami. Martin explained that it took three loads to cover Skelton's 166-acre cotton field. He admitted that he had sprayed flagmen, but he knew of no problems with their health. When he flew near the gin, he explained, he was doing trimming work around the edge of the field, and on his approach he flew under the power lines and very close to the gin and nearby house. As he focused on the fields and adjoining hazards, Martin insisted, he did not recall seeing any people on the platform. While he admitted that at some time that day he no doubt had passed over the platform, "I can say I didn't spray the platform."[21] Focusing on the wires and buildings and the edge of the cotton field, people were outside his visual equation. In that sense, Martin was a typical crop-duster pilot, part of a tradition that by 1956 spanned thirty-five years.

When the gaunt and emaciated Charles Lawler took the witness stand, he personified a workman stripped of his skill and pride. He had made a good living and reared three children to adulthood, and he was providing a good home for his two younger daughters. From 1950 to 1956, he had worked for the Marie Gin Company, spending the off season "on the road doing gin work." His productive working career ended abruptly. No longer able to work, Lawler first directed Irene Lawler in managing the Marie Gin and then accepted what he judged a job offered out of sympathy, but ultimately he became too ill even to pretend to work. When V. A. Johnson fired him in the fall of 1957, Lawler lost his housing, and his household broke up. A man who had once scaled ladders and worked up high, he became dizzy simply negotiating steps. Pesticide exposure left him emotionally unstable—he was unable to control his crying—and he was helpless without oxygen. In the judgmental culture of the Mississippi Delta, he had fallen from a productive workman to a welfare recipient.[22]

Lawler testified that he had lived in Elaine, Arkansas, until 1951, when he moved his family to the Marie community. In the spring of 1956, in order to work for another gin, he had a medical examination. Dr. Rozelle Hahn of Indianola gave him a clean bill of health. "I hadn't lost ten days work from well before 1950 until then," he stated, referring to the poisoning. He recalled that on the afternoon of August 15, Tracey Skelton met V. A. Johnson "and we seen them walking and checking the field just north of the house and the fields around and near the gin building." Skelton told him they were going to poison the next morning, and on the morning of August 16 went by the gin to remind him. At that point, Lawler told him that he was in the midst of welding steel and that unless he completed the job the steel would warp.[23]

Lawler then gave his account of the critical incident. He was working with Johnny B. McCaleb and George Washington, both experienced gin men. Jessie Lee Pam had gone to the store for drinks when the plane came over. "I was hunkered down welding with my helmet on," Lawler testified, "trying to get through so the crew could get away from there if the poison began to come in on us." On the stand, Lawler pointed out the geography of the cotton fields and the gin structures and indicated that the plane was traveling east to west. "Johnny B. said the wings barely missed the tip of the gin shed," Lawler testified. "That is when he opened the valve and when I got my poison." He was in the midst of the steel beams with his welding rods, hammer, and electrodes. When McCaleb saw the plane, he shouted, "Look out, Mr. Lawler," and stepped under the edge of the shelter. Lawler's back was to the plane; when he heard McCaleb shout, he had no time to react before he was covered with a mixture of endrin, malathion, and xylene. The spray covered his back, neck, and head, "and it came through the top of my helmet." The poison "stifled me," Lawler continued, and "it was about three minutes before I could get a breath to let me relax or straighten up after it was over."[24]

After he got his breath, Lawler went home. He could taste the poison, he remembered, and was nauseated and nervous; he had a headache. Nonetheless he returned to the gin, completed the welding job, and supervised the workers during the afternoon. At suppertime he drank only a cup of coffee. During the night, he woke up sick. "I went to the bathroom and vomited quite a bit," he testified. He noticed that "the calf of my legs and my arms and muscles were jerking, and in my back." The next morning he returned to the gin, but at 11:00 A.M. he went home with a high fever and soon lapsed into a coma. Irene Lawler called Dr. A. A. Aden, who rushed him to the hospital, where he spent three days. Townsend and Hulen led Lawler through his travails after returning home, which culminated in his inability to work. When asked if he had worked since 1957, he replied, "Not a day or not an hour."[25]

Lawler's son, Charles S. Lawler, explained that prior to August 16, 1956, his father was in good health, that there were two children aged nine and thirteen living at home, and that he and his sister presently supported their father. "His lungs are in such condition," Lawler explained about his father, "that at times he chokes completely down, and without oxygen, he can't continue to breathe." His father, he testified, "is in bed 98% of the time."[26] Despite her vital involvement in the incident, Irene Lawler was not called to testify.

Johnnie B. McCaleb was the only surviving eyewitness to the poisoning, as George Washington had passed away. The trial transcript listed the thirty-five-year-old McCaleb as "(Colored)," just as the other witnesses were listed "(White)." McCaleb reviewed the facts of the morning of August 16. He had not paid any attention to the crop duster until it suddenly appeared. "The airplane dived over the road down below the light wire post; it come from across the dredge ditch and come over going west," he recalled. He had time to warn Lawler and to step under the shed. The spray, he observed, "looked like milk, liquid or something other like that." He likened the density to "a little sprinkle." After helping Lawler to his feet, he wiped the residue from Lawler's face. He recalled seeing V. A. Johnson and Tracey Skelton riding along the edge of the cotton field the day before. Under cross-examination, Forrest Cooper sought to confuse McCaleb and fault his sensitivity to white supremacy. In his questions about Skelton's decision to poison his crops, Cooper asked, "He was discussing with you colored boys his personal affairs?" McCaleb clarified that Skelton was making a general statement to everyone there, including Lawler. Cooper asked, "Mr. Johnson was not out there very often, was he?" McCaleb surprised him by replying, "Every week or couple of weeks." Cooper asked how McCaleb could so clearly recall that on August 15, Johnson had inspected the cotton fields and Skelton had brought cotton bolls to the gin to examine them. "I just remember," he answered. In their notes, Townsend and Hulen wrote that although Johnny McCaleb was "brow beaten" by Forrest Cooper, his "testimony was not shaken."[27]

Something extraordinary happened to Charles Lawler on August 16. It is possible that the hands who mixed the final batch of endrin and malathion got it wrong. The welder's helmet might have funneled chemicals into Lawler's lungs. His welding posture might have exposed him to the biplane's unique spray swirl. Stop-action photographs show that spray coils around a biplane's right wing, pushed by the propeller slipstream and by wingtip vortices. The swirling spray might well have focused on Lawler. Because Lawler worked around cotton lint and smoked cigarettes, he might have been highly susceptible to the poison mixture. None of the expert witnesses were curious about such questions.

Defense attorney Forrest Cooper used expert testimony to buttress conventional wisdom that pesticides used in farming were not harmful to humans. One of his witnesses, Dr. Mitchell R. Zavon, worked at the Kettering Laboratory in Cincinnati and practiced occupational medicine. Born in Woodhaven, New York, in 1923, Zavon had studied agriculture at Cornell University before taking a medical degree from Harvard University in 1949. He had worked several years for the PHS and was certified in occupational medicine. Ambitious and confident, he eagerly assumed leadership and consulting positions. He boldly introduced himself to the court as a medical college professor, Cincinnati's director of occupational health, a medical consultant for the agricultural chemical division of Shell Chemical Corporation, a consultant for the PHS's National Poison Control Center Clearing House, chairman of the toxicology committee of the National Agricultural Chemicals Association, and director of the Cincinnati Poison Control Center. When Cooper asked Zavon to delineate the geographical extent of his pesticide expertise, the scientist brazenly replied, "It includes the whole world."[28] Mitchell Zavon wore many hats, a role that would eventually raise conflict of interest questions.

For the purposes of the Lawler trial, Zavon considered himself an expert on agricultural chemicals, including endrin and malathion. He had earlier spent a week and a half in the delta investigating agricultural practices. When asked if humans standing near a cotton field would be harmed if sprayed with endrin and malathion, Zavon declared, "There would be no hazard to the man; I would not like to have chickens running around in the field." Cooper asked if Zavon in all of his experience had ever known

Mitchell Zavon.
Compliments of Mitchell Zavon.

"any person who has been permanently injured while he was standing out in the open while exposed to this solution for only a short time." Zavon replied, "I have seen none, and there had been no reports of such." Even in an enclosed space, Zavon asserted, endrin and malathion had two possible results on humans: "They either die right then and there or get well—there is nothing in between." For survivors, he proclaimed, there were "no after effects." With a vial of endrin, Zavon demonstrated the small amount present in a mixture applied in a hundred-square-foot area.[29]

Elizabeth Hulen had objected several times to what she considered Forrest Cooper's leading questions. In her cross-examination, she led Zavon through complex explanations of the toxicity of chlorinated hydrocarbons and organophosphates, pushing him to admit that xylene, malathion, and endrin were dangerous poisons. She initiated a series of questions relating to the effects of malathion poisoning, which resulted in a long discussion of cholinesterase, an enzyme involved in transmitting nerve impulses. Hulen used the *Cecil-Loeb Textbook of Medicine* as her guide, and Zavon took issue with it. "This is absolutely incorrect," he replied to one of Hulen's queries, but he admitted that *Cecil-Loeb* was a "recognized textbook." Zavon also disagreed with Wayland J. Hayes Jr.'s claims that animals might have convulsions up to 120 days after exposure to endrin, although he favorably cited another of Hayes's experiments. "Dr. Hayes has reported in the scientific literature in experiments with human volunteers in the Atlanta Penitentiary there were no observable effects and no effects were observed as a result of feeding DDT over a many months period, there were absolutely no effects on liver function or any other observable or laboratory results," he proclaimed.[30]

The Hayes prisoner study had become an instant classic that was favorably cited by such chemical apologists as Zavon and members of the ARS staff. To Rachel Carson, Clarence Cottam, and other environmentalists, however, the study epitomized the lengths that scientists would go to justify the use of pesticides. In October 1958, Carson wrote to nature writer Edwin Way Teale that one of her "delights" in the book that she was working on "will be to take apart Dr. Wayland Hayes' much cited feeding experiment on '51' volunteers. It was '51' for only the first day of the experiment; thereafter the experimentees rapidly lost their taste for DDT and drifted away until only a mere handful finished the course of poisoned meals." Nor had Hayes followed up on the prisoners' health, she pointed out. Ultimately, Carson did not include her critique of Hayes's study in *Silent Spring.* When Hayes moved to Vanderbilt University, he continued his interest in human testing. In 1974 he proposed to the Environmental Protection Agency that Veterans' Administration hospital patients and volunteers at the state prison be used in studies of the effects of herbicides in humans, as well as a drug that would combat alcoholism.[31]

Wayland J. Hayes Jr. always found something good in pesticides, and his work was cited approvingly by chemical advocates. "The public health benefits from the use of agricultural chemicals," he wrote in a 1954 article, "should always be kept in mind in considering the potential hazards involved in the use of these materials." When people were poisoned, he suggested, it was their own fault: "The cases of poisoning were all caused by excessive exposure to the insecticides and were frequently associated with gross carelessness on the part of workers." The "newer agricultural chemicals," he found, had an excellent safety record. Drugs caused more household poisonings than pesticides. "Once an accident has occurred," he complained, "the injury to public relations has been done." After two children were poisoned by the organophosphate TEPP, he fretted that "the Communicable Disease Center has received correspondence inquiring why the use of such a poisonous material as TEPP is permitted." Magazine articles, he continued, unnecessarily alarmed people. "In the mature, basically agricultural community," he concluded, "the necessity for using agricultural chemicals is accepted, as is the potential danger which some of these materials bring to those who use them." In a 1961 article, Hayes condemned people who read antipesticide articles and "attribute their vague, recurrent, or chronic disease to minimal or even nonexistent exposure to chemicals." When discussing reduced cholinesterase levels in workers, he suggested that a person who suffered only mild symptoms "should be allowed to continue his work." If cholinesterase levels decreased again, he suggested, "one must assume that the subject is unable to follow safety regulations and he must be removed from contact with toxic materials." According to Hayes, imagined, not actual, poisoning was the problem. When a worker actually became ill from organophosphate poisoning, it was clearly the worker's fault. In Hayes's estimation, chemicals were benign.[32]

Elizabeth Hulen turned to labels used on pesticide bags, inquiring if Mitchell Zavon had any part in writing the label for endrin. The label, he explained, had been written before he began consulting for Shell in 1957. He agreed that endrin labels should be marked "poison" with a skull and crossbones. Hulen then read the cautionary parts of the label, including the statement, "Do not apply or allow to drift to areas occupied by unprotected humans or beneficial animals" and the danger warning about absorption through the skin, inhalation, or swallowing; she noted that the label explicitly stated, "Spray mist extremely hazardous." Zavon agreed that the label was necessary and reflected the dangers of the poison. Hulen next turned to malathion. Zavon explained that cholinesterase allowed nerves to connect properly and that organophosphates interfered with that function. The first symptoms, he explained, were not unlike endrin poisoning: stomachache, headache, and nausea. Hulen then explored solvents with Zavon. Discussing exposure to kerosene, Zavon explained that the primary danger was vomiting, which pushed the toxic material into the lungs. Hulen suggested that the results were often labeled bronchial pneumonia, but Zavon opined that "chemical pneumonia" was a more accurate term. Hulen explained to the court that xylene was the solvent in endrin, but that the solvent mixed with malathion was not specified on the label. "It could have been benzene or kerosene or any of these," she stressed.

Zavon admitted that a person exposed to a small dose of xylene "might develop headache, giddiness, faintness, perhaps nausea, and go through all the stages of anesthetic down to unconsciousness."[33]

Zavon had visited the Mississippi Delta for a week and a half "in order to observe what was going on and how they did these operations—how they grew the cotton, how they sprayed it, how they mixed the materials in the formulating plant, what safety precautions were taken, how airplane spraying was done, how they loaded the planes, how the fields were sprayed, insofar as we could, the total agricultural operation and everything that went into it." He visited the staffs of "one or two clinics" and "one or two health offices" and discussed "specific cases of intoxication from agricultural chemicals with them." His evaluation of safety practices ranged from "minimum to nonexistent" for field workers to "pretty good" for duster operations. Zavon had never known a flagman to wear any protective equipment, whether in Mississippi, California, "or any place else." Hulen read from the *Safety Manual for Aerial Applicators,* which warned pilots to take special precautions to avoid spraying houses, people, or livestock. "The organic phosphate insecticides," the manual observed, "are the outgrowth of a search that was carried on in Germany during World War II for new chemical warfare agents." Zavon would not budge from his insistence that pesticides used according to directions were not harmful to humans.[34]

Hulen and Zavon battled over Charles Lawler's exposure to the toxins. Zavon denied that the welder's helmet acted as a chimney that channeled the pesticides onto Lawler's face and into his lungs. Johnny McCaleb saw the whitish chemical residue on Lawler's face, Zavon explained, because it was a hot day and only "a little of this material" could cause "a whitish discoloration." Zavon insisted that Lawler could easily have raised his helmet. Hulen pointed out that his hands were occupied with welding tools, he was surrounded by steel, and he was choking. When Hulen mentioned pesticide research at the Mayo Clinic, Zavon replied, "There has been some trash come out of the Mayo Clinic in regard to some of these materials." He obviously distrusted Dr. Malcolm Hargraves's research and findings. Elizabeth Hulen's relentless cross-examination displayed her legal skill and her mastery of pesticide issues. Zavon had his hands full as he danced around the implications of her interrogation.[35]

Expert witness Dr. Griffith E. Quinby did not appear in the Indianola courtroom but submitted an affidavit. He had published a study of chemical health hazards in the Mississippi Delta based on a ten-week investigation he made from June 16 through August 28, 1957. The State Agricultural Extension Service sponsored the study, which was cosponsored largely by agribusiness stakeholders: the State Plant Board, the Delta Branch Experiment Station, the Delta Council, the USDA, the Mississippi State Aeronautical Board, the Mississippi State Board of Health, and the Mississippi State Medical Association. Born in Kentucky in 1914, Quinby studied medicine at the University of Louisville and took a M.A. in public health at the Johns Hopkins University in 1946. From 1952 to 1954 he served as medical officer at the Communicable Disease Center (CDC) in Savannah, Georgia, where, he explained, he was "in charge of animal laboratories and studying all records of poisoning from pesticides collected from all over the United States." It is likely that he knew Wayland J. Hayes Jr., who was then working at the Savannah facility. In July 1954 he moved to the Wenatchee Field Station of the CDC in Washington, where he continued his pesticide studies. Confident and articulate, Quinby had traveled internationally investigating the health effects of chemicals, including "organic phosphorus chemical warfare agents similar to many of our present day insecticides." His written responses had the ring of a confident doctor clearing up misconceptions and lecturing the unlearned.[36]

After explaining the chemical composition of both endrin and malathion, Quinby observed that in 1952 malathion was sprayed in orchards in the Northwest; it was used experimentally in Mississippi in 1956 and widely during his 1957 tour. Farmers had applied endrin in the delta since 1951. Before his investigation of the delta, Quinby had coauthored a study of the effects of TEPP, another organophosphate pesticide, on pilots' vision; he had also coauthored a study of parathion poisoning among West Coast field workers.[37]

Since Quinby left Mississippi in late August, he was not there when *Business Week* correspondent John Donaldson asked about parathion poisoning in Glen Allan, but several people had mentioned to him that workers in Glen Allan were being sickened. Quinby had briefly visited Glen Allan to pursue Dr. Mary Hogan's reports of poisoning. "I talked with the physician concerned first on the phone and later went to the area to investigate." During this time, Charles Lawler was visiting Dr. Hogan. Quinby conducted cholinesterase tests but did not publicize the results. He made no mention of seeing patients waiting in the clinic yard, in Dr. Hogan's words, "coughing up or spitting blood." Neither in Glen Allan nor in any part of the delta, Quinby asserted, had he discovered any person injured by malathion or endrin. "I was not able to find convincing cases of poisoning in anyone handling malathion or endrin unless they drank it." Despite a number of publicized cases of poisoning and other medical evidence, Quinby recalled only one death from malathion poisoning, "the case of a South Carolina boy of low mentality walking behind a mule-drawn sprayer directly in the path of the sprayer and wet plants."[38]

Quinby testified that he had discussed Lawler's illness with Dr. Aden but did not mention the case in his report "because I gave him my requested opinion that this case was not due to the alleged insecticide exposure and that a far more satisfactory diagnosis surely could be found." To Elizabeth Hulen's long and complex hypothetical question about the context of Lawler's exposure to pesticides, Quinby predicted that other than smarting eyes, "no other ill consequences should have been expected to be due to this insecticide mixture." Quinby ignored the question of potentiation, the stepped-up danger that sometimes arose when chemicals were mixed together. None of Lawler's symptoms could be the result of pesticide poisoning, he avowed. "A complete clinical study of this patient," he reiterated, "should have produced a satisfactory diagnosis other than alleged insecticide poisoning."[39]

Quinby's dogmatic deposition raised serious questions not only about his analysis of pesticide poisoning symptoms but also about his research methodology. Although he spent ten weeks in the delta researching pesticide poisoning, he could not find time to examine any of the hundreds of Dr. Hogan's poisoning patients around Glen Allan. Nor did he publicize the results of his cholinesterase tests on workers near the town. Dr. Hogan had successfully treated her patients with atropine, another indication that they were poisoned by parathion. Even more puzzling, Quinby discussed the Charles Lawler poisoning case with Dr. Aden but never examined the patient. He nonetheless dismissed Lawler's symptoms as unrelated to pesticide poisoning, although they were the classic symptoms of endrin, parathion, and xylene poisoning.

Since Dr. Aden had first treated Lawler, his testimony for the defense was crucial. Aden had moved to Indianola in 1942 after studying medicine at both the University of Mississippi and Emory University. He had treated Charles Lawler for the better part of a year after the poisoning incident. The defense produced x-rays taken by Dr. Rozelle Hahn before Lawler's poisoning that showed traces of emphysema, suggesting a preexisting illness. Aden analyzed the x-rays and observed emphysema. In her testimony, Hahn explained that she had treated Lawler for influenza, and from the x-rays she had determined that he had "a minimum amount of emphysema," about what she would expect from a fifty-year-old man who worked in a cotton gin. In Hahn's opinion, Lawler's current health problems resulted from pesticide poisoning. In the mid-1950s, while practicing in Indianola, she had treated two patients poisoned by organophosphates. One was a crop duster who "had some foul up in his system, and he got a good amount in the process of dusting." She had administered atropine, and both patients made a full recovery.[40]

Aden recalled that when he had treated Lawler after the accident, Lawler had complained of "coughing and pain in his chest," and Aden had diagnosed pneumonia. "If Mr. Lawler mentioned anything about poison," he testified, "I don't remember it." When asked if he had treated Lawler for poisoning, he reiterated that he had treated him for "virus pneumonia"; later, he stated, "He got no treatment for insecticide poison." When Lawler went to his office in December, Aden "thought he possibly had had a heart attack." Looking at x-rays from that incident, Aden admitted, "I did make a mistake on the picture. He did have some pneumothorax at that time." Aden went through Lawler's files step by step, explaining the treatment in detail. He testified that it was only in mid-January 1957—five months after Lawler had been taken ill—that poisoning came up as a cause of Lawler's distress. Under Elizabeth Hulen's cross-examination, Aden remembered little of what happened on August 17, 1956. At the time, Aden confessed, he did not know if Lawler "had pneumonia or whether he was having a heart attack." Nor could he recall if Lawler complained of vomiting, coughing, fever, dizziness, or headache. When Hulen asked, "You diagnosed it as a virus pneumonia, therefore, you gave him no atropine?" "No," Aden testified, responding "no" again when asked if he had checked Lawler's cholinesterase level. Aden recalled only symptoms connected with chest pains. He did remember that Lawler was nervous. "I have been out there to see him," Aden recalled, "and he would be upset and crying and we would have to give him something to quiet him down." When Lawler's health did not improve, Aden sent him to Dr. Watts R. Webb at University Hospital in Jackson.[41]

With Aden on the witness stand, Dr. Watts R. Webb's five-page letter regarding his examination of Charles Lawler on May 29, 1957, was introduced into the evidence. "The patient was agitated with crying at one point during the initial interview," Webb observed. Lawler's lung capacity was diminished by one-third, "demonstrating a fairly marked degree of respiratory disability." He had lost some eighteen pounds, and his appetite had fallen off. "He has likewise become extremely 'nervous,' easily upset and is unable to tolerate even low heights off the ground where formerly he worked 30 to 40 feet above the ground without difficulty," Webb noted. Despite his coughing and breathing difficulties, Lawler had continued to smoke a pack of cigarettes a day after the poisoning, but at the time of the examination he had stopped smoking. Webb prescribed a regimen of medication and observed that Lawler's wheezing decreased, his coughing slacked off, and he breathed more easily. Working in a dusty and lint-filled cotton gin and smoking cigarettes contributed to Lawler's emphysema and bronchitis, Webb concluded. "There is no doubt that this was tremendously aggravated by the cotton spray," he decided. "From his description of the episode it is rather amazing in fact that he lived through it." Webb wrote that if Lawler refrained from smoking, avoided dusty situations, and used "broncho-dilators and antihistaminics . . . he may show some improvement over his present status." With Webb's letter before him in court, Aden still claimed that Lawler suffered no immediate effects from the poisoning. He agreed that as Lawler's physician he had completed a Mississippi Workmen's Compensation Commission report on June 25, 1957. He described Lawler's injury as "irritation to the trachia-bronchial tree resulting in acute bronchitis and pneumonia resulting in spontaneous pneumothorax."[42] Aden was either incredibly absentminded or purposely vague in his testimony, and he distanced himself from Lawler's pesticide exposure.

The defense, intent on proving that pesticides were harmless, called Cecil Black, the county agricultural agent for Sunflower County. Reflecting on his nineteen years' experience, Black stated that being sprayed by pesticides "is a natural occurrence." He had been sprayed, he admitted, "where a plane passed over and sprinkled me, and all the flagmen were sprinkled, and observers—anybody in the field naturally is going to get sprinkled." He had never seen any ill effects from such exposure. Pascol Townsend then asked him to read the endrin label. Black agreed that the precautions outlined on pesticide labels were prudent, but added, "in actual practice, they just don't work out." He had never seen the *Safety Manual for Aerial Applicators.* Chemicals were a vital part of delta agriculture, and Black contended that the current safety regulations were sufficient, even if they were ill-observed. G. G. Ames, a crop duster who lived in Drew, admitted to spraying flagmen and discussed the dangers—including chemical hazards—that pilots faced. The defense paraded local farmers and another aerial applicator before the jury to testify that they knew of no one harmed by endrin and malathion. Another dozen witnesses were waiting to testify, including Dr. Marvin Merkel, when the defense rested.[43]

To counter the claims that people exposed to chemicals never suffered adverse effects from them, Pascol Townsend and Elizabeth Hulen put Dr. Mary Elizabeth Hogan on the stand. Her experience with pesticide poisoning countered the defense witnesses who claimed chemicals were benign. Charles Lawler had visited her Glen Allan office in August 1957, she recalled, "when I began to get illnesses that I felt were due to cotton poison." She answered a series of questions revealing that, in her opinion, Lawler's illness was neither chronic bronchitis nor chronic emphysema. When she stated that in August 1957 she had treated patients suffering from "cotton poisoning," Forrest Cooper objected that her testimony was not rebuttal. Judge Greene sustained the objection. When Hulen asked Hogan to summarize Griffith Quinby's report, Cooper again objected, although he permitted the entire report to be placed in the record. Hulen suggested that jurors would not read all of the report, but Cooper's objection was sustained. Pascol Townsend pointed out that Cooper had only revealed this report several weeks before, and he felt that a competent witness should analyze it. Hogan testified that when Quinby had visited Glen Allan, he had conducted a cholinesterase test but did not test for xylene or even mention endrin. "I tried to get him to go up to see some of my patients in Linden plantation," she continued, "but he said he had to go back to the State of Washington, that he had been in the Delta all summer, and that he had to return to Washington and didn't have time."[44] Hogan was obviously still smarting over planters, public health officials, and Quinby dismissing her diagnosis of chemical poisoning. Cooper's objection prevented her from relating her experiences or explaining how she disagreed with Quinby's conclusions.

Quinby's report was placed in the record as an exhibit. Working out of the Delta Branch Experiment Station in Stoneville, Quinby had visited fields, tested workers, solicited material from delta physicians, and examined hospital records. He had found 100 cases of suspected pesticide poisoning, "including 11 episodes involving either an airplane crash or a disturbance of a pilot's performance." He had examined 92 of these cases, ultimately dismissing 78 as "some other cause." He had conducted no exposure studies of malathion, although he wrote that "alleged poisonings were investigated." In 1957 planters increasingly moved from malathion to parathion as the pesticide of choice. Despite warnings about pesticide toxicity, planters, workers, pilots, and loaders were careless. Many pilots loaded their own planes and carelessly discarded pesticide containers; they eschewed the need for regular cholinesterase tests. Planters and pesticide sales agents "concealed or minimized" pesticide dangers. Despite such negligence, Quinby found only one fatality between 1954 and 1957, a child who swallowed endrin. He found two nonfatal parathion poisonings in formulating plants, three "mild" cases among loaders, and three cases among pilots.[45]

Quinby's report purported to take "epidemics during the summer of 1957" seriously. "Only one physician attributed any substantial number of these illnesses to poisoning," he wrote, not mentioning Dr. Hogan by name; but he acknowledged that "many people were concerned over the claims of this one physician that the epidemics were due to poisoning, especially when widespread newspaper and radio publicity was given to this possibility." Citing results from cholinesterase tests and other data, Quinby ruled out poisoning. In addition, he mentioned a team from the Mississippi State Board of Health that collected specimens. "Neither of these surveys showed any indication of poisoning," he asserted. Quinby did not mention the poisoning symptoms that Dr. Hogan reported, nor did he note her successful treatment of the "respiratory infections" with atropine.[46]

Of the eleven "airplane crashes or disturbances," Quinby considered three as possible organophosphate poisonings. One pilot crashed in 1956 after "erratic flying" while applying malathion dust; he had applied parathion "some weeks previously." Another had difficulty landing on August 12, 1957, "because he was dizzy, unable to think clearly, and had double vision. He vomited while still in the air." There was parathion seepage into the cockpit, and he had been exposed to the pesticide several days earlier. By shutting one eye, the pilot managed to focus enough to land successfully. After applying parathion for several days, a pilot spilled guthion on his hands on August 20, 1957, but washed it off quickly. After he took off, "he had visual disturbances, headache, tightness in the chest, abdominal cramps, nausea, vomiting and weakness." He landed safely, if unsteadily, and was given atropine. "All of these circumstances pointed to poisoning," Quinby admitted. He also performed residue studies for parathion and guthion and checked workmen who entered cotton fields after pesticide applications. According to Quinby, organophosphates were for the most part benign.[47]

As with his deposition, Quinby's article on pesticide use in the Mississippi Delta minimized dangers from organophosphate pesticides. He eagerly looked for other causes of sickness both with Charles Lawler and with the "epidemic" that swept the delta in the summer of 1957. Much like John Martin, who focused on dangerous obstacles and did not see people below his crop duster, Griffith Quinby focused on positive aspects of poisons and was blind to their obvious dangers. As a public health physician, however, Quinby's oversights were disturbing. Mitchell Zavon, Griffith Quinby, and Wayland Hayes Jr. were a powerful troika that spread calming words about pesticides.

Both sides made their closing arguments. After four days of testimony and an hour and a half of closing arguments, it took the jury twenty minutes to return a verdict for the defendants on March 19, 1960. In early April, Townsend and Hulen appealed the case to the Mississippi Supreme Court. Undisputed testimony, they argued, showed that the chemicals in question were "poisonous" and potentially dangerous and that J. L. Turk, through his pilot, was "negligent in releasing the liquid insecticide on the gin platform and on plaintiff." As a result of being sprayed with a mixture of endrin, malathion, and the xylene solvent, the plaintiff had suffered an acute illness unrelated to any preexisting health condition. Johnson and Skelton, as joint adventurers, were both liable for Lawler's damages. Forrest Cooper answered, citing expert witness testimony.[48]

The Mississippi Supreme Court handed down its decision on May 22, 1961. The court did not accept Forrest Cooper's claim that chemicals were not dangerous, writing, "It is undisputed that, if a person receives an excessive amount of these chemicals, they can be highly toxic and dangerous to human life." There was no evidence that disputed the claims of Lawler and McCaleb that the plane had sprayed the platform. The court took particular exception to Dr. A. A. Aden's testimony. It reviewed his claim that pesticide poisoning was first mentioned five months after Lawler was hospitalized. On the witness stand, Aden had denied that he had prescribed atropine, but "the hospital record, however, reflects that on August 17 he prescribed atropine, the recognized antidote for chemical poisoning." The trial court should have allowed the rebuttal testimony of Dr. Mary Hogan, the court ruled: "The trial court erroneously sustained an objection to this, since the inquiry was proper on rebuttal, and would have tended to contradict defendants' witnesses." The court accepted the jury verdict that the landlord-tenant relationship between V. A. Johnson and Tracey Skelton was not a joint adventure. "In summary," the court ruled, "the judgment in favor of Turk and Skelton is reversed, because it is against the overwhelming weight of the evidence, and the cause is remanded for a new trial as to them."[49]

Even before the decision, the Delta Council's B. F. Smith mobilized Mississippi farm leaders when he learned that the appeal was going against the defendants. A group of six attorneys met with Smith "to consider action that might be taken." They decided to file a petition with the supreme court "requesting permission to appear as *amicus curiae* and for leave to file a brief on suggestion of error. This petition was made in the name of Delta Council and the Mississippi Farm Bureau Federation." Chemicals had become an integral part of the state's economy, the petition argued, "and the welfare and prosperity of our people are dependent upon the continued ability of farmers of this State to produce their crops effectively." Claiming to speak for all farmers in the state, the petition contended that "this case unless modified or restricted will have a serious and adverse effect on the agricultural economy of this State."[50]

As farm organizations realized, cotton had become one of the most chemical-dependent crops, and pesticides had thus become indispensable to cotton farmers. In 1956, the year that Charles Lawler was poisoned, a USDA bulletin listed twenty-nine insect pests that threatened cotton crops. It recommended ten formulations for boll weevils: aldrin, BHC, calcium arsenate, chlordane, chlorthion, dieldrin, endrin, heptachlor, methyl parathion, and toxaphene. Malathion was not recommended for boll weevils in 1956, although it was for several other cotton pests. By 1959, guthion, malathion, and sevin had been added to the boll weevil list. Farmers were always eager to expand their knowledge of the constantly evolving roster of chemicals, asking county agents and chemical salesmen for advice on formulations that would be most effective in protecting their crops. The 1959 USDA pamphlet recommended that workers stay out of the fields for fifteen days after methyl parathion application and a lesser time for other chemicals. "Guthion, methyl parathion, and parathion are very poisonous acutely," the 1959 booklet warned. "Use great caution in applying."[51]

Responding to the reversed ruling's potential impact on chemical use, Forrest Cooper prepared a suggestion of error. He labeled the supreme court's pending decision "the worst blow to the cotton farmers of Mississippi that they have received in many decades." The *Lawler* case had grave implications. "Any man can now claim that he breathed some of the dust or spray and no matter how unjust or unfair his claim may be, the farmer is liable unless he has an eye witness or other direct evidence to dispute the claim," Cooper wrote. Chemicals were not "ultrahazardous," he insisted. He praised the expert witnesses who had denied that chemicals caused Lawler's sickness. At every opportunity Cooper linked Lawler's case to Johnnie McCaleb, the sole African American to take the witness stand. "No witness except the plaintiff and his negro helper testified that the plane put out an excessive amount," he argued. After discussing testimony that the spray nozzles were set and could not be changed, Cooper wrote, "Now, that testimony is not contradicted by Lawler himself and his negro helper. . . . Where is there any evidence in this entire record other than that of the plaintiff and his negro helper that he received an excessive amount?" Cooper observed that "the record is full of the testimony of doctors and other experts casting serious doubt on the statement of this plaintiff and his negro assistant." Cooper enlarged upon this idea by making a general observation: "If the by-stander says he was 'saturated' and has a negro helper to testify for him and the farmer shows that the equipment used made such saturation impossible, even though a jury of twelve good and lawful men find he was not saturated or hurt, the farmer is helpless."[52] The implication was that the word of neither Charles Lawler nor Johnnie McCaleb could stand up to expert testimony.

When the Mississippi Supreme Court turned down the petition of the Delta Council and Farm Bureau and overruled Forrest Cooper's suggestion of error, B. F. Smith grumbled to Farm Bureau president Boswell Stevens that the agribusiness strategy had suffered a clear setback. Smith suggested that the situation "appears to demand legislative action early in 1962 to remove the threat of numerous law suits that might develop." Stevens agreed and vowed to find someone immediately to draft a bill "that would adequately protect our people who, of necessity, will be using the various chemicals, and the service of those people who are in the business of poisoning crops for various reasons."[53]

Lawler v. *Skelton* was never retried. Whether McCaleb was unavailable, Lawler was too ill to testify, or Townsend and Hulen were simply convinced that no delta jury would convict Skelton and Turk, the lawyers gave up on a jury trial. By early 1962, they were in negotiations with Forrest Cooper over a settlement.[54]

As the final settlement process continued, Irene Lawler appealed directly to Elizabeth Hulen. Throughout the ordeal, she had remained a stoical supporter of her husband. She nursed him, helped him work when he was able, and from time to time corresponded with his attorneys. She thanked Hulen and Townsend for their work on the case but was disappointed that they would "realize no more out if it than actual expenses." She believed that V. A. Johnson's power had prevented a successful resolution to her husband's case. "I am not a lawyer and know so little about law," she admitted, "but I am asking you if there isn't a court where a person can get justice against a man of Mr. Johnson's political & financial power." She reminded Hulen that Johnson opposed Lawler in his workman's compensation claim. "Watching Mr. Lawler slowly dying from this poison, the financial strain and effects on our two young daughters," she concluded, "has almost been more than I could stand." Hulen forwarded the letter to Townsend with the message, "I wish you would try to explain to her." By June the papers were filed, and Tracey Skelton and J. L. Turk settled for $4,599.00.[55] Charles Lawler lived for five more years, passing away on June 10, 1967.

Even as the *Lawler* case was being settled in June 1962, the *New Yorker* carried the first installment of Rachel Carson's *Silent Spring.* In retrospect, *Silent Spring* has had an eerily similar effect on environmental questions as the 1954 *Brown* v. *Board of Education* decision on segregation. While both *Silent Spring* and *Brown* were monumental and set in motion historical currents that endure, neither accomplished its ultimate goal. After *Brown,* the South, if not the nation, is today vastly different from the segregated and hateful place it was in the 1950s and 1960s. By the same token, *Silent Spring* created movements that helped halt the use of the most persistent pesticides. Yet the volume of chemicals, not to mention genetically modified organisms, used in the United States today is far greater than when *Silent Spring* was published. Despite the changes since *Brown,* the residue of racism persists with the tenacity of DDT. Indeed, as so brilliantly portrayed in Connie Curry's book, *Silver Rights,* and her film, *The Intolerable Burden,* Sunflower County schools are still segregated, the white pupils now in private schools and blacks in the white-abandoned public schools.

The agribusiness community saw in Rachel Carson's book an imminent threat to its livelihood, just as the delta's white elite saw in the *Brown* decision a challenge to segregation and worker control. Both chemicals and segregation served—and continue to serve—the white delta elite. In Indianola, as in other delta towns, chemicals were linked to agricultural prosperity and thus to the general economic welfare. While the health complaints of black workers could be ignored or dismissed, Charles Lawler's poisoning was more complex. A skilled worker and family man, Lawler was part of the white community. Yet the community—and the jury—rallied around V. A. Johnson, Tracey Skelton, and J. L. Turk. Pascol Townsend and Elizabeth Hulen could not sway a rural community of stakeholders who condoned pesticides and thus supported the planters. Chemicals sometimes poisoned minds, bodies, and souls. Yet Elizabeth Hulen, Dr. Rozelle Hahn, and Dr. Mary Hogan challenged the controlling ideology, foreshadowing women's enthusiastic response to *Silent Spring.* When Carson warned of invisible dangers such as radioactivity and chemicals, atomic testing and widespread application of chemicals had long been accepted as benign. Radiation experts dismissed health complaints from people downwind from nuclear facilities, and chemical advocates claimed that *Silent Spring* was flawed and launched a massive attack to discredit it. Faced with a growing environmental movement, the chemical industry became even more ruthless after the publication of *Silent Spring.* Still, agribusiness was unable to discredit it. *Silent Spring* emerged as the bible for the environmental movement, in large part because it resonated with thousands of pesticide incidents in homes and fields and in communities across the country.

3
THE RISE OF SKEPTICS

Last week some 2500 acres of farmland were sprayed with this compound [dieldrin] near Bolivar, Tennessee, and I would like to venture a guess that the beetle faired out much better than any other type of life in the area, for immediately after the treatment were found 19 dead rabbits, 22 quail, many song birds, fish in stock ponds, not to mention frogs, field mice, and snakes, and all in only 350 acres of the areas sprayed.

John R. Priora, April 23, 1961

As the *Lawler* case made apparent, rural people were often exposed to pesticides; and as poisoning cases spread across rural and urban America, suspicion intensified that chemicals endangered not only fish and wildlife, but also humans. Enthusiastic advertising obfuscated warnings on pesticide labels and in cautionary newspaper and periodical stories. In addition to pesticides used on farms and gardens and in homes, the Agricultural Research Service (ARS) indiscriminately drenched large areas of the country with pesticides in efforts to control gypsy moths, Dutch elm bark beetles, fire ants, and other pests. By the mid-1950s, pesticide spray drifted across the country. And as poisoning cases multiplied, more people questioned the efficacy of unbridled spraying.

Long before *Silent Spring* placed the pesticide issue in context, farmers, crop dusters, field workers, and sportsmen had learned from experience that some chemicals were toxic to humans, domestic animals, fish, and wildlife. The pilots who flew the Stearmans, Pipers, and Travel-Airs were on the front line of pesticide knowledge. They handled chemicals daily, as did their loaders, and their lives depended upon understanding the toxicity of their cargoes. Along with tractors, mechanical harvesters, and other implements, airplanes played a crucial role in transforming rural life. In the South, aerial crop dusting emerged in the early 1920s, at about the same time that tractors began appearing with some regularity on southern farms. While many farmers dusted by hand or used mule-drawn apparatuses, larger operators took advantage of the aerial application of pesticides. When farmers seized on the powerful synthetic pesticides that emerged after World War II, dusting technology evolved to support the new formulations. Pilots and loaders learned from experience about chemical dangers, and although many were macho in their disregard for safety, most clearly respected pesticide toxicity.

Applying pesticides from the air began in August 1921, when the Ohio Department of Agriculture successfully dispensed powdered lead arsenate from a Curtis JN-6H to combat the catalpa sphinx moth. Southern crop dusting—centered at Scott Field, near Tallulah, Louisiana—quickly followed. World War I aircraft could easily be retrofitted with a hopper, a crank, and dispenser, and in August 1922, pilots demonstrated dusting techniques to farmers and entomologists in the cotton country. Aerial application of pesticides had clear advantages over hand dusting or even mule-mounted sprayers; despite the marginal effectiveness of calcium arsenate, Paris green, and nicotine sulphate, crop dusting became a thriving business. Dusting companies worked closely with entomologists and with large cotton plantations, such as the Delta and Pine Land Company. Capitalizing on the demand for dusters, in 1925 the Huff-Daland Airplane Company centered its operations in Monroe, Louisiana. By 1929, the Delta Air Service controlled the assets of Huff-Daland Dusters, and Delta pushed into Texas, Florida, and Mexico. (Delta Air Service ultimately evolved into Delta Airlines.) The early days of crop dusting featured daring pilots, excellent mechanics, and studious entomologists—a combination of talents not uncommon in the South. The early history of southern crop dusting thus refutes the stereotype of a backward and mechanically unskilled working class. Indeed, the South had always harbored a skilled labor force, one that had tinkered with plows and wagons, run steam engines, manufactured and repaired cotton gins, engineered and maintained locomotives, operated sawmills, and serviced automobiles. In skill and temperament, pilots and airplane mechanics were close kin to stock-car drivers and mechanics. Pilots and race drivers possessed enormous skill and toughness, combined with a disregard for dangers that would frighten or immobilize most people.[1]

Discussing the old days with oral historian Lu Ann Jones near Clarksdale, Mississippi, in 1987, pilot Cotton Carnahan recalled that before World War II he and his fellow pilots used primarily "calcium arsenic and cal-nic—that was nicotine mixed with calcium arsenic, straight sulfur and old Paris green." Arsenic, he told Jones, "didn't kill too many bugs.

Killed a lot of cows and horses and mules." Carnahan was born in Mooringsport, Louisiana, in 1920, and he got hooked on flying in 1937 while in college. By 1940 he was an instructor and also was dusting with a 4000 Model Travel-Air. Like most young pilots, he liked to fly low, "knocking the leaves off the cotton." Much the same combination of factors that propelled the post–World War I expansion of aerial pesticide application recurred in 1945. Experienced pilots, surplus airplanes, newly developed synthetic pesticides, and obstinate pests reconfigured prewar operations. Cotton Carnahan bought a used Stearman for $182, paid $500 for a hopper and $750 for a spray rig, and he was in business. Pilots often made their own hoppers out of plywood or galvanized tin, Carnahan recalled: "We soldered the seams or glued the plywood. It wasn't very sophisticated." Pilot Jack Shannon, putting that in perspective, allowed that a new Grumman crop duster in 1987 cost around $145,000, and a turbine might go for up to $300,000.[2]

Jack Shannon was born in Sac City, Iowa, in 1927; his folks farmed livestock and corn. He served in the Air Force during World War II and moved to Mississippi in 1950. He started dusting with a J-3 Cub, moved to a Stearman, and bought his own plane in 1952. Shannon worked in the Texas rice country as well as in the delta. In the 1950s, he recalled, the dusting season lasted only three months; by the 1980s it extended to nine months. Only in the dead of winter did duster pilots rest. Chemical companies used the slack months to host fish fries and introduce new chemicals to farmers and dusters. When crunch time came, planters expected dusters to appear as if by magic and the chemicals they sprayed to produce instant kills. Planters blamed pilots if insects or weeds survived, and Shannon admitted that sometimes pilots made mistakes.[3]

Duster pilots needed a Federal Aviation Administration Agricultural Operating Certificate, and their planes had to meet minimum federal requirements. Yet even the best of pilots could not completely control toxic drift. When 2,4-D was sprayed on narrow-leafed rice, corn, or wheat crops to kill weeds, it easily drifted to nearby broad-leafed fields. A number of complaints came from cotton farms that bordered Texas, Louisiana, and Arkansas rice fields. In June 1948 the U.S. Civil Aeronautics Administration banned 2,4-D dust, but liquid application was hardly an improvement. According to USDA bulletins, even in calm conditions 2,4-D in liquid form drifted 1,350 feet when sprayed from an altitude of 20 feet and 550 feet when released from 10 feet. Chemicals followed air currents created by the propeller slipstream, which rotated counterclockwise toward the

Cotton Carnahan and Chuck Thresto, Clarksdale, Mississippi, October 8, 1987.
Photo by Laurie Minor Penland, LM 97-16537. National Museum of American History.

right wingtip, producing vortices that followed airflow over and along the wings. Chemicals rode this turbulence, traveling on breezes or exploiting atmospheric conditions, often drifting far from the intended field. A three-mile-per-hour breeze could carry a droplet eight miles when applied from ten feet above the field. The USDA received a number of complaints in

1948, including a case when drift from 10,500 pounds of 2,4-D sprayed in two Louisiana parishes left a wide swath of damage to vulnerable cotton fields, including 297 acres that were completely destroyed and 471 acres that were heavily damaged.[4]

Pilots always tested the wind before applying pesticides, and the wind was most quiet in the early morning and late afternoon. The duster's day started before light, and often pilots worked from the farm that was being sprayed. "We'd fly off of turnrows or just little old pastures, anything," Cotton Carnahan recalled. The farmer furnished the poison and the help to mix and load it. Just after World War II, engines were not reliable; Carnahan remembered nine forced landings in one day. In the postwar era, he added, "every foot of cultivated land in this Delta was cotton. Some places you could run five or six miles in one pass." When the Rural Electrification Administration came to the delta, every tenant house had electricity, so there were many wires to dodge. "In a three-mile run," he recalled, "I would either jump or go under 18 wires."[5]

Duster pilots often discussed chemicals, and some of the preorganic compounds were dangerous. Black Annie was a pre–World War II cotton defoliant composed of 16 percent non-acid-forming nitrogen fertilizer; it was produced by the American Cyanamid Corporation. It was sooty black and had a disturbing side effect. "Any alcohol'd make you fire up," Shannon laughed. "Turn red as Cotton's cap." Carnahan added, "Your fingernails would turn purple. You could feel your heart beating on your fingernails." This naturally led to a discussion of drinking. "Back in the old days," Carnahan recalled, "if you saw a duster pilot he was nine times out of ten an alcoholic to go along with it." Some would carry the bottle with them while flying. There were accidents "caused from a person's head all full of cobwebs."[6]

R. C. Colvin grew up in Carroll County, Mississippi, and in 1940 he enrolled in the Civil Pilot Training program. He joined the Royal Air Force and spent the first years of World War II in England and Egypt. In 1945 he settled in Greenwood, Mississippi, bought a Stearman, and "started into the crop dusting business." Before the war, he explained, "crop dusters had a pretty bad name for quite a while because they put a swath here and put a swath there and they're through. Hadn't done a job worth a durn. They'd collect their money and gone. That's what you call 'fly-by-night' pilots." By the early 1950s, Colvin had fourteen pilots working for him. In the years just after the war, he recalled, pilots were "hard-living, hard-drinking, hard-womanizing." Still, Colvin stressed that pilots concentrated on their work. "You are doing dangerous work," he explained to Lu Ann Jones in October 1987. "You are low. You got lots of high lines. You got houses. You got a lot of obstructions." While spraying in Louisiana, he once encountered "a hidden high line." He admitted that he had not cased the field carefully, recalling, "I woke up with that line right in front of me and tried to go through it." The propeller cut the line, but it wrapped around the struts and caused him to crash. He spent a few days in the hospital. After admitting that some people were allergic to pesticides, he declared that handling chemicals was not a big problem: "Rachel Carson, she just messed up the whole works and made you scared to touch anything." Loader boys, he explained, "walk around out in the loading pit out there in bare feet and poison, stuff like that."[7]

To orient the pilot, a flagman often stood at the end of the field marking each swath. Usually African Americans, flagmen received the brunt of the pesticides, but any impact on their health has gone unrecorded. Pilots and observers admitted that flagmen were customarily exposed to chemicals. The Sunflower County, Mississippi, county agent admitted that he had been exposed to spray, "and all the flagmen are sprinkled, and observers—anybody in the field naturally is going to get sprinkled." There were other dangers as well. As Nathaniel Huff flagged on the R. N. Aldridge plantation near Arcola, Mississippi, in September 1957, the airplane struck him and put him in the hospital with a concussion.[8]

The men Lu Ann Jones interviewed in 1987 had witnessed not only major transformations in chemical toxicity but also in rural life. When Carnahan began dusting after World War II, plantations were dominated by tenant and sharecropper labor. "Little old shotgun tenant house, with an outhouse out behind it," he remembered. There were often people in the fields. "But back in the old days, we'd just spray or dust right on over 'em. It never did hurt any of 'em that I know of. Never hurt any of us that I know of." Then came tractors, picking machines, and ever more chemicals, and the hands moved off. Government programs were supposed to help small farmers, Carnahan recalled. "Well, they didn't turn out that way. It just helped 'em out of farming into a job in town. Most government programs don't work like they're supposed to." Jim Lawhorn likened the dynamics of the crop dusting business to that of agriculture in general. The independent owners were selling out because running a business was too expensive. "That's the way American agriculture is going—it's become a volume business," he observed. "Us small independents, the ag pilot and farmers, we're being forced out by the marketplace."[9] Crop dusting ultimately became a cog of agribusiness, vulnerable to the same economic squeezes that affected farmers.

As Donovan Webster researched his *New Yorker* article, which appeared in 1991, duster pilots continually mentioned Melvin Hayes, who had flown B-17s in World War II and then piloted dusters for thirty years. Webster found him in the deep woods near the Mississippi River, where he lived a self-sufficient and reclusive life. Hayes sneered at dangers from pesticides and boasted of dusting with sulfur dust, DDT, and benzene hexachloride (which burned him when it came into the cockpit). He deplored Rachel Carson and the banning of DDT. "I've never been sick a day in my life," Hayes boasted. "I've never been to a doctor, except when I had my Army physicals. Never had a cold, or anything." He had been exposed to most of the agricultural chemicals used since World War II, he said, "and the only health effect they ever gave me was that one day I spilled some DDT on my boot and it cured my athlete's foot." Hayes personified the macho pilot. For a while he flew secret missions for the U.S. government, and he also smuggled drugs. "After a while," he philosophized, "flying ag isn't interesting enough anymore, so you got to find other ways to have fun." He was self-sufficient, stubborn, and opinionated, and he had little patience with bleeding hearts. An "old boy" delivered beer, ice, and radio batteries to him. "Here in the bottoms, I can do what I want. I don't have any neighbors. I know where the deer are. I know where the fish are," he bragged.[10] Hayes's attitude exemplified not only that of many dusters but also that of farmers and the public at large. Most people did not have immediate reactions to pesticides, and they felt confident that the government would warn them of dangerous poisons. Indeed, they found it hard to believe that toxic poisons would be allowed on the market.

Unlike Melvin Hayes, most duster pilots admitted to the dangers of chemicals. Born in 1922, Chuck Thresto grew up in Texas and joined the U.S. Army Air Corps in 1942. A month after he left the corps in 1949, he began dusting with a J-3 Cub in the Rio Grande Valley. "And we knew that some would bite you quicker than others would," Thresto observed of synthetic chemicals. In retrospect, he wondered if regulators had overestimated the dangers of chemicals, "because we have since then done everything but bathe in the stuff. And a few have even taken inadvertent baths, I think." His friend Cotton Carnahan, however, was quite leery of parathion. "It would affect your vision first," he recalled. "It appeared that you were looking in the wrong end of a set of binoculars." Although Thresto had never been poisoned, he had respect for pesticide toxicity and admitted "that any of them, under the proper conditions, are toxic and life-threatening in the strengths that we handle them." Toxaphene would drift, Carnahan remembered. "It'll go ten miles to get in a fish pond, and kill every fish in there."[11]

Jim Lawhorn, who flew out of West Helena, Arkansas, described to Donovan Webster the dangers of flying: trees, wires, other airplanes, and finally, chemicals. "As an ag pilot, you've always got to be mindful of the toxicity of what you're laying out," Lawhorn warned. His hopper was full of methyl parathion mixed with other chemicals, and it could kill catfish, eat the paint off of cars, and harm people. "I mean, you get a drop of that undiluted methyl parathion into an open cut on your hand, and, man, you die."[12]

Pilots often presented a macho and irreverent image to the world. Jim Lawhorn, a Vietnam veteran, was drawn to flying as a young man. Experience counted a great deal in dusting, he told Webster. "If you're lucky enough to live through your first few crashes," he judged, "then your odds of staying alive rise significantly." Webster realized that duster pilots treated crashing as almost part of their job, much as stock-car drivers accept the fact that crashes are sometimes unavoidable. Old-timers called themselves duster pilots or ag pilots, although in time a more fancy name emerged. Carnahan cleared up the terminology. "They always said duster pilots sounded too common, or something," he joked. "So they started calling themselves aerial applicators."[13] These pilots realized that their lives depended upon care in handling pesticides and that their business thrived on accuracy of application.

The pilots who flew for the Agricultural Research Service (ARS) control projects, on the other hand, ignored property lines and displayed little of the finesse shown by private dusters. As Rachel Carson labored on *Silent Spring* in the late 1950s and early 1960s, many people were first introduced to pesticide toxicity by ARS control planes, which fogged their land and homes and killed birds, fish, and wildlife in addition to the intended target. While agricultural dusters focused on fields in rural areas, control pilots swept indiscriminately over houses, barns, pastures, and sometimes small towns, seemingly unmindful of collateral damage. Environmentalists monitored government insect-control programs and posed tough questions about the increasing use of poisons and their detrimental effects on wildlife.

ARS control programs generated strong opposition. Neal Landy complained to his Pennsylvania senator in May 1957 that "it is a question about the invasion of private property without permission either by the owner or by court order." As a World War II veteran, Landy reminded the USDA's True Morse that DDT used in the exigencies of wartime saved lives. "But this is not a matter of life or death," he judged of the gypsy moth control program, "but one of bureaucratic expediency." Roger Hinds, who brought a suit for fifteen Long Island residents against the USDA, wrote in July 1957 that the plaintiffs "resent having their persons and lands doused from the air by low-flying planes, without their consent and without any showing that either their persons or their lands are contaminated, with the Gypsy Moth, the current target of the 3,000,000 acre, $5,000,000 project of the Department of Agriculture and the less-enlightened chemical manufacturers."[14] While USDA bureaucrats dutifully responded to such criticism, they nonetheless defended the use of chlorinated hydrocarbons and expanded control programs across the country.

Some complaints prompted ARS investigations. In May 1957, when Spring Valley, New York, farm owner Hanes Fried complained to Congresswoman Katherine St. George about DDT spraying, the ARS's V. E. Weyl visited Fried's farm. He was dismayed that "all in the area seem to be organic farmer enthusiasts." The farm's dairyman explained that the local health inspector allowed him to market unpasteurized milk, but he was concerned that the milk might contain traces of DDT. Weyl explained the control program and reassured the dairyman that there would be no more spraying in the area. "I also told him that the material would break down rapidly in the sun and dissipate," he reported, although chlorinated hydrocarbons do not break down rapidly, if at all. Farm manager Ralph Courtney complained of bird deaths as well as indiscriminate spraying without the owner's permission. Weyl related that "a Miss Burns thrust her way into the room and conversation." She was "a bird enthusiast" and "an organic garden enthusiast although she smoked cigarettes," he observed with displeasure. She presented Weyl with four dead birds. The area had been sprayed on May 1, 2, and 6, Weyl acknowledged in his report, and "the material was properly applied."[15] Weyl's visit not only revealed ARS skepticism about organic farmers and bird enthusiasts but also the willful spread of inaccurate information. By 1957, pesticide residues were present in most of the country's milk and meat supply, and scientists and many laymen understood that chlorinated hydrocarbons were persistent and did not, as Weyl said, break down rapidly.

The ARS often blurred fact and fiction in justifying its control programs. Although fire ants were not economic pests—that is, they did not seriously interfere with farming—ARS publicists and other supporters transmogrified the insect into a monster that devoured quail, calves, crops, and the tender flesh of children. In fact, fire ants thrived on a diet of insects but stung viciously when bothered. With little research into the life cycle of fire ants or the effects of powerful chlorinated hydrocarbons such as heptachlor, dieldrin, and aldrin on nontarget insects, wildlife, domestic animals, or humans, the ARS pledged not just to control fire ants but to eradicate them. By 1967, the ARS had sprayed 45 million acres with chlorinated hydrocarbons to control fire ants. Ultimately the ARS overextended itself and created a backlash among conservationists, sportsmen, and farmers, who objected to invasive spraying and the resulting collateral damage.[16]

It was in creating and defending this needless and doomed eradication project that the ARS revealed its willingness to falsify data, intimidate opponents, and disregard scientific evidence. As the fire ant campaign gained momentum late in 1957, Clarence Cottam emerged as its major critic. Born in Utah, Cottam graduated from the University of Utah in 1923; he earned a M.S. degree in 1926 from Brigham Young University and a Ph. D. from George Washington University in 1936. Cottam worked for twenty-five years for the U.S. Biological Survey and its successor, the U.S. Fish and Wildlife Service (FWS). In July 1954 he resigned in protest when President Dwight D. Eisenhower appointed Douglas McKay as Secretary of the Interior. When McKay fired FWS director Albert M. Day, Rachel Carson charged that McKay and his business-oriented appointees would "return us to the dark ages of unrestrained exploitation and destruction." After a year as dean of Brigham Young University's College of Biology and Agriculture, Cottam became head of the Welder Wildlife Refuge near Corpus Christi, Texas. The National Audubon Society awarded him its ninth Audubon Medal in 1961, and he was honored by the National Wildlife Foundation in 1964.[17]

Clarence Cottam combined disappointment, outrage, and common sense in his attacks on the fire ant project. In a November 1958 speech, he ridiculed the notion that fire ants could be eradicated and speculated that no "competent entomologist" in the USDA "sincerely and honestly believes that eradication is within the realm of possibility." When ARS field workers found no wildlife damage after spraying, Cottam branded their reports "misleading" and "prepared either in alarming ignorance of the facts or with a lack of intellectual integrity somewhere along the line." While he accepted "reasonable controls where there is a proved need," he insisted that "there is a moral obligation to see that other economic and cultural resources are not unnecessarily damaged in the process."[18]

Prefacing its boilerplate replies with feigned concern for fish and wildlife, the ARS's defensive rhetorical barrage claimed to embrace "such values as human health, domestic animals, and fish and wildlife." It falsely claimed close cooperation with the FWS, the Department of the Interior, the Public Health Service, and state agencies. It boasted that not one farmer on the 500,000 acres sprayed for fire ants by December 1958 had complained. A skeptical Cottam forwarded a letter from a field worker who reported that the ARS stifled complaints and that the author of a USDA report purposely distorted evidence in a "deliberate and malicious effort on the part of the USDA to mislead the people by misuse of facts."[19] The massive spraying for fire ants, mosquitos, beetles, and other pests alarmed many people who simply did not want toxic residue falling on their houses, land, and animals.

Disgusted by ARS lies and threats to fire field workers who complained, Cottam charged on February 15, 1960, that he had never witnessed a control program "that has been so poorly handled nor one that has exhibited such a lack of intellectual honestly as this one." He lambasted one ARS bureaucrat for self-serving errors in a letter to a congressman, ridiculed the lack of prespray research, scorned the ARS's decision to begin heptachlor application at two pounds per acre only to discover that one and a quarter pounds was sufficient, and charged that the ARS "rigged" its wildlife reports. "Miller," he wrote to his old friend Miller Shurtleff at the USDA, "haven't *some* of your control workers, who I don't believe are recognized authorities in wildlife, taken unto themselves an air of omniscience and omnipotence that is just a little repugnant?"[20]

Having begun his career at the USDA as a messenger, Miller Shurtleff had known Cottam for twenty-five years, ever since Cottam was assistant director of the Biological Survey (then part of the USDA). Shurtleff was strangely reluctant to confront the ARS about Cottam's "pretty serious" charges. "In past correspondence I have tried not to answer Clarence in too much detail," he confessed, "because the whole situation is rather delicate."[21] Shurtleff's cautious replies to Cottam and his reluctance to confront the ARS suggests the presence of a veiled but substantial bureaucratic muscle within the ARS. No one in the USDA, it seemed, was willing to call the ARS into account.

The ARS derived its power from operating at the intersection of pesticide regulation and use, between approving the chemical industry's formulations and ensuring that foods were not contaminated. Chemical industry representatives roamed ARS hallways, consulting with staff about labels, residues, control projects, research, and other issues. The ARS had become not only a pesticide clearinghouse but also a sprawling bureaucracy. It had troops in the field, planes to apply pesticides from the air, and a cadre of generals in Washington who led the war on insects. ARS leaders enforced discipline and demanded strict allegiance to its programs. Clarence Cottam looked upon the ARS with dismay, even disgust, for he recalled the time before the triumph of synthetic pesticides, when USDA entomologists were well known for sound scholarship. In 1961, the Federal Pest Control Review Board attempted to address some of the worst pesticide abuses. It gave the Department of Interior and the FDA a formal role in pesticide policy and encouraged coordination among all agencies that dealt with pesticides. Still, the USDA jealously guarded pesticide policy.[22]

As Cottam pointed out, the compulsion to spray overcame ARS concerns for fish and wildlife and for ethical standards in scholarship. With its persistent claims that it was concerned about wildlife, that it cooperated with other agencies, and that pesticides were benign, the ARS emerged as both a major pesticide booster and consumer. Evidence of wildlife losses, domestic animal deaths, and even problems with human health failed to impede the ARS's compulsion to spray. It alternately promised and postponed research, and it used public relations to deal with the consequences.

Cottam's accusations are substantiated in ARS files, which bulge with reports of and complaints about wildlife destruction and questions regarding human health. Obviously within the ARS there was an ideology that condoned falsifying data in order to continue the dubious fire ant program and other control projects. Only rarely does reflection or introspection appear in ARS memos and letters. Because the ARS portrayed itself as a science-driven bureau that was utilizing the latest weapons in its protection of food and fiber, it enjoyed immunity from close oversight. Staffers hid behind white lab coats and white lies and countered criticism by claiming that their efforts protected the country's food supply and helped feed a starving world. In retrospect, the fire ant campaign was a doomed battle to eradicate an insect of marginal economic importance. After three years of massive spraying across the South, there were more fire ants than when the program started, but there had been enormous collateral damage. Had the control program scenario been scripted by Hollywood, ambitious and brilliant fire ants would have cleverly provoked the spraying program as a way to destroy their enemies, expand their range, build resistance to pesticides, and conquer the world. ARS bureaucrats would have appeared as hubristic, bumbling dolts and pseudo-scientists at the service of the chemical industry.

While the toxic intrusion of control projects alerted many people to pesticide problems, farmers encountered dangers close at hand. They were often puzzled and alarmed when they lost livestock after their land had been sprayed. W. H. Welch Jr., who ran a tree nursery and cattle operation in Romeo, Michigan, complained in June 1960 that the Detroit Edison Company sprayed his pasture with the herbicide 2,4,5-T without obtaining his permission. He immediately moved his cattle from the treated area. Four months later, he reported, a 500-pound steer fell dead, three nursing cows "have gone to skin and bones," and another cow was unable to "get with calf." Welch asked the secretary of agriculture if the meat from the dead cow, which was in his freezer, was "fit for human consumption." He also asked what effect 2,4,5-T had on cattle. In its July 1960 reply, the ARS cited a 1953 Michigan State University study that concluded 2,4,5-T "seemingly has no harmful effects on livestock" and an ARS study that "found 2,4,5-T to be of very low toxicity for livestock." The reply concluded that 2,4,5-T "could not have caused the sudden death of your 500 pound steer in November." The ARS suggested that Welch ask his local veterinarian about the frozen meat and also have him examine the sick cows.[23]

No accurate count of livestock deaths exists, but court cases and ARS investigations suggest a high toll. To keep weeds down along highways, Louisiana Department of Transportation (DOT) workers sprayed monosodium acid methanearsonate (MSMA) along Louisiana Highway 15 south of Vidalia six times over three years. The spray truck operator insisted that he had duly shut off the spray fifty yards before and after the cattle underpass at the S. L. Winston Jr. plantation. Still, arsenic built up in the standing water in the culvert that ran underneath the highway and poisoned four bulls. Veterinarian Joe Price Lancaster deduced that the dead bulls had drunk from the arsenic-contaminated standing water in the poorly drained culvert; other bulls had diarrhea, a symptom of arsenic poisoning. The Louisiana DOT sprayed some one thousand linear miles along highways each year to control Johnson grass and weeds. Spraying, it calculated, was cheaper than mowing.[24]

Forreston, Texas, farmer C. S. Baker complained to then-Senator Lyndon Johnson in October 1960 that three of his cows had been poisoned. A neighbor had sprayed defoliants on his cotton, and the airplane had "circled over my place." Prior to the spraying, he had been offered $250 each for the cows and had spent $39 attempting to save one of them. "There have been several cattle killed in our county by the dfolant [*sic*] they are using," he charged. He asked Johnson for advice on how to get fair compensation for his dead livestock, since his entreaties to his neighbor had been unsuccessful. J. V. Beck, an official of the Izaak Walton League who lived in Grand Island, Nebraska, complained to Congressman John D. Dingell in August 1961 that "aerial sprayers are spreading among other chemicals, parathion far and wide over the corn fields, even in the face of repeated warnings from the entomologists of the Ag College about the danger of this poison." He learned that "some cattle have been killed up near Bartlett." When the USDA replied to Dingell's query, it admitted that it was investigating "the killing of 10 heifers as a result of an insecticide spray." It also admitted that each year there were accidents in pesticide use.[25]

As suggested by its reaction to Dr. Mary Elizabeth Hogan's experience in Glen Allan, Mississippi, the ARS became adept at rapid deployment of "experts" to defuse charges that chemicals damaged fish, wildlife, livestock, or humans. Time and time again, ARS representatives or staff from other federal agencies rushed to an area and cleared pesticides as the cause of health problems. In some cases, no doubt, there were other causes at work, but in many instances—too many to be a coincidence—wildlife and cattle losses accompanied spray campaigns. In April 1962, even as the *Lawler* case headed toward settlement, ARS pesticide teams near Monroe, Louisiana, sprayed for fire ants directly across a bayou from a pasture with a hundred head of cattle. About a week later, thirty-one of the cattle died. Their symptoms—loss of appetite, diarrhea, and loss of coordination—indicated poisoning. The Louisiana State Diagnostic Laboratory announced heptachlor poisoning but retracted that verdict on May 3. The ARS, meanwhile, sent R. D. Radeleff from its Animal Disease and Parasite Research facility in Kerrville, Texas, to investigate. Since there were still active fire ant mounds on the farm, Radeleff deduced that heptachlor had not drifted over the pasture and concluded that "it was obvious that the cattle had not been poisoned by heptachlor." He found an old hanger on the property, in one corner of which he discovered "a bin of cotton seed, over which were spread sacks of salt, mineral mix, tools and what-have-you." He speculated that local authorities would "find the cause of poisoning to be associated with the contents of the old hangar, probably to be due to a failure to provide salt, then a gorging by the cattle, followed by the losses." Radeleff also reported "a general hysteria in Monroe concerning heptachlor." He was distressed that two veterinarians "appear to be making diagnoses by telephone, considering that a plane flying over a premise, followed by death of an animal, is evidence enough for a diagnosis of heptachlor poisoning." He criticized the Louisiana Diagnostic Laboratory for originally citing heptachlor in the cattle deaths. "More unfortunate," he complained, "is the damage done to the Fire Ant control program by such irresponsible statements as are being made by veterinarians and physicians in Monroe." Investigations wasted control resources, he mused, and the downtime costs "must have been considerably more than the value of the 31 cattle Mr. Watson so unfortunately lost."[26] Radeleff's convoluted hanger theory, his impatience with local authorities, and his admission that there were other cattle deaths in the area suggested a larger problem than discarded material in a hanger.

Radeleff had worked for the old Bureau of Animal Industry, and after World War II he had conducted toxicity studies on livestock exposed to DDT. In 1947, the Entomology Research Division (at that time the Bureau of Entomology and Plant Quarantine) set up its livestock insects laboratory at Kerrville, Texas. There Radeleff worked on insecticide residues in milk and meat. In a November 1962 address, he boasted of how few cases of livestock poisoning occurred when treatment was carried out *"in strict compliance with label directions."* Yet he was concerned that even when animals were apparently poisoned, "neither veterinarians nor physicians can confirm a diagnosis of insecticide poisoning." They relied upon "signs, symptoms, history and so on, but cannot by any laboratory method prove their conclusion." This caused serious problems when control programs coincided with livestock or wildlife deaths, he admitted before a panel of the Federal Council of Science and Technology in November 1962. Entomologists "tend to think of poisoning as an immediate, rather easily recognized condition," he continued, but over the years animals developed problems weeks or even months after being near pesticide applications. "Because of the time delay," he explained, "most, if not all of these have been attributed to factors other than pesticides." Then he made a glaring admission: "I am not entirely certain that we were correct in doing this." Pesticides were clearly at work in cattle deaths. He cited several cases of cattle poisoning that he had closely monitored that suggested the lingering effects of poisons. "We have been extremely fortunate, during 15 years, in being able to apply educated guess work when experimental data were unavailable," he confessed. "How long these educated guesses may be utilized in protecting the livestock grower and the public is anyone's guess."[27]

For some people, first-hand observation fostered skepticism about bureaucratic pronouncements on pesticide use, often prompting letters of protest. Women, in particular, took an active role in environmental issues. In February 1960, Mrs. Thomas R. Dillon complained to her senator that during the previous spring, birds, squirrels, and other wildlife in northern Illinois towns were "slaughtered from DDT spraying of Elm trees for Dutch Elm disease." After the spraying, dead wildlife was "picked up by the sacks full and the stunned people wanted to know who had done it." From her reading, the Wheaton, Illinois, resident continued, she had learned that such spray campaigns "not only are wiping out our wildlife, but are very seriously threatening the health of the nation." Mrs. Dillon dismissed the usual USDA disclaimer that pesticides increased food production to feed a starving world. In fact, she concluded, "billions are spent for crop storage and taking cropland out of production." In April she complained that in Detroit heptachlor granules were "dropped on residences, parks, farms, schools, factories." Sportsmen, she continued, had no way of knowing how much residue was in their kills. From Glen Ellyn, Illinois, Mrs. Arthur M. Jens Jr. complained that to combat Japanese beetles, "highly toxic heptachlor was sprayed in granular form from a B25 twin engined bomber, and every square inch was covered." Agricultural authorities, she fumed, refused to comment on dead wildlife.[28]

Testifying in April 1964 before Senator Abraham Ribicoff's subcommittee of the Committee on Government Operations, Ruth Graham Desmond, president of the Federation of Homemakers, charged that in Michigan dieldrin had fallen on schoolchildren. Desmond also related the story of a family that lived beside a commercial orchard. Spray drifted over the family's patio, exposing the mother and her child to "dangerous pesticides." Five of the woman's horses died that year. The family sold the home and sued for damages, but it took the woman four years to regain her health. "During the trial," she revealed, "the orchard owner testified he used in early spring lead arsenate in solution followed by DDT, dieldrin, parathion, Phosdrin, Systox, and Guthion." If pesticides had such a dire effect on healthy people, Desmond asked, what would it do to those who were ill?[29]

Local control projects also elicited vociferous complaints. In April 1961, O. D. Brattan of Memphis protested to Senator Estes Kefauver about the white-fringed beetle control program. "I don't know how serious the insects are that they are trying to kill," he judged, "but I am sure they cannot spray the whole world and kill them all." Brattan was "an ardent hunter" and was alarmed by a newspaper clipping that discussed "game birds killed" by pesticides. John R. Priora reinforced Brattan's concern, sending Kefauver an account of wildlife kills in Hardeman County, Tennessee. Priora reported in April 1961 that 2,500 acres had been sprayed with dieldrin, "and I would like to venture a guess that the beetle faired out much better than any other type of life in the area." He listed the kills: "19 dead rabbits, 22 quail, many song birds, fish in stock pond, not to mention frogs, field mice, and snakes." While the damage to wildlife was obvious, Priora warned that pesticides might also threaten "to a small degree our own health for a certain amount of this chemical will reach our tables along with the food that was grown on the treated farmland." He urged Kefauver to stop the spray campaign, "for I believe the wildlife of the United States is of much more importance than the control of one beetle at the present time." Since the country had "nothing resembling a food shortage . . . now or foreseeable in the next decade, I can see no reason why this spraying of farmlands all over the country cannot be stopped immediately." He enclosed a *Memphis Press-Scimitar* article by Paul Fairleigh that reported massive wildlife kills in Hardeman County. In the USDA's reply to Kefauver, it denied wildlife kills. "According to our records to date," Assistant Secretary Frank J. Welch wrote in May 1961, "thousands of acres have been treated for this insect in 11 States and in none of the areas treated has there been any indication of appreciable losses of wildlife."[30]

Some people understood the contradictions inherent in an agricultural policy that on the one hand created a vast surplus and on the other hand encouraged greater productivity through science and technology. Early in 1962, Alice Durand of Puposky, Minnesota, criticized extension agents who, instead of teaching farmers husbandry, in essence worked as chemical salesmen. "These boys, fresh out of college and indoctrinated by the or shall I say with the help of the Chemical Companies go out & sell products to farmers which in the long run prove dangerous and harmful to him and everyone else," she fumed. "In addition to that the taxpayer is paying their salary," while "they are actually salesmen for the chemical companies and are subsidizing them." Indeed, in February 1960 the National Association of County Agricultural Agents issued a press release announcing a Shell Chemical Company contest "to help spread information about successful ideas for pest control programs because of their importance to modern farming." The press release boasted that "the county agent has played a vital role in introducing new chemicals and new practices to the growers in his county." As Durand suggested, it was frequently difficult to determine where county agents ceased to be public servants and became chemical salesmen.[31]

Although the USDA's claims to feed the world gave it widespread legitimacy, some critics uncovered contradictions in its myriad programs. In August 1962, shortly after *Silent Spring* appeared in the *New Yorker*, Onekama, Michigan, resident Barbara Bolling notified Kenneth Birkhead, assistant to the secretary of agriculture, that the USDA "suffers from gargantuan obsolescence." She objected to the subsidies paid to large commodity producers and to the indiscriminate spraying of pesticides. The USDA, she pointed out, subsidized "wheat and other grain which we can't use and must store, and spends great sums on rural agents to serve non-existent farmers, on automatized farms who are used by banks and other absentee farms owners for what?" Funds saved by ending subsidies and spray campaigns could be turned to purifying the water and used for research on biological control, she suggested.[32]

Birkhead's defensive reply relied upon the false analogy of automobiles, which people depended upon but could live without. "Yet, automobiles are deadly weapons when misused," he cautioned, stressing that thousands of people were killed each year on highways. Laws controlled dangerous automobiles, he wrote. "The same is true about insecticides." Birkhead's self-serving analogy attempted to obfuscate the truth about the massive and indiscriminate spray campaigns, chemical residues in foods, wildlife destruction, and accidental human deaths. He spent several paragraphs attacking Rachel Carson's data in *Silent Spring* and defending chemical companies. "We are not losing our birds, animals and wildlife," he declared. "The balance of nature always changes and I am not sure that I would like to see a return to the days when men cowered in a cave, fought off monsters, and was lucky if he reached age 30."[33] In Birkhead's historical analysis, eliminating synthetic chemicals that had been on the market a mere fifteen years would instantly catapult society back to the dark ages. False analogies and specious historical claims typified USDA statements and those of corporate spokesmen and defenders who rode the popular scientific wave. If problems arose, the cliche ran, then science could solve them.

In the seventeen years between World War II and the publication of *Silent Spring* in 1962, most people embraced the promise of a chemical future; others were ambivalent. Complaints bubbled up all over the country from people whose personal experience or research provoked questions about pesticide safety. Rachel Carson almost magically articulated the country's hunger for pesticide information. That her book found such favor meant that it addressed deep and widespread apprehension. That it terrified chemical companies meant that it was on target.

Born in 1907 in Springdale, Pennsylvania, Carson was a quiet and reserved child who loved books and nature. Her first story, "A Battle in the Clouds," won the *St. Nicholas* silver badge for excellence in prose; she was eleven years old at the time. Family financial difficulties only accentuated her shyness. At the Pennsylvania College for Women, she made lasting friends and fell under the spell of several strong women professors, including biology professor Mary Scott Skinker. Carson enrolled in biology classes at the Johns Hopkins University in 1929 but left in 1934 because of financial difficulties. Two years later she took a job with the U.S. Bureau of Fisheries, and in 1942 she was promoted to assistant aquatic biologist and assigned to the Fish and Wildlife headquarters at the Department of the Interior. Carson worked as an information specialist, a position that combined her writing talent and her scientific training. In this position she became acquainted with many of the leading fish and wildlife authors; she also gained a knowledge of current research and the latest science. In 1945 Clarence Cottam and Elmer Higgins's disturbing reports on health effects from DDT came to her attention. She unsuccessfully proposed an article on DDT to *Reader's Digest.* Carson brilliantly translated complex scientific research into everyday language. In 1951 she published *The Sea Around Us* to high critical acclaim, winning the National Book Award for nonfiction. The following year, she resigned from her government position to pursue her writing.[34]

Gradually Carson was drawn into the debate over pesticides, especially questions regarding fish and wildlife destruction and human health. "In the process of her research," Carson's biographer Linda Lear wrote, "she established a remarkable network of scholars in many fields all over the world and created an alliance of scientists, naturalists, journalists, and activists committed to helping her document a spectrum of environmental abuses." Her contacts from her years in the federal bureaucracy opened doors, and concerned workers eagerly informed her of environmental abuses. These contacts and her own wide reading allowed her to synthesize the literature on pesticides.[35]

Silent Spring resonated with Americans' increasing concern about pesticides and the environment. The book became a best seller and made Rachel Carson a household name. Carson, her allies, and her growing number of disciples alarmed chemical companies, their bureaucratic allies, farmers, and pesticide advocates throughout the country. Both her supporters and her opponents understood the significance of *Silent Spring*. In Linda Lear's estimation, the book was "a fundamental social critique of a gospel of technological progress. Carson had attacked the integrity of the scientific establishment, its moral leadership, and its direction of society." When CBS announced plans to televise "The Silent Spring of Rachel Carson," the chemical industry exerted enormous pressure to have the program cancelled, and three of its five sponsors withdrew two days before the April 13 telecast. While Carson comported herself with calm dignity on camera, Dr. Robert White-Stevens arched his dark brows and prophesied, "If man were to faithfully follow the teachings of Miss Carson, we would return to the Dark Ages, and the insects and diseases and vermin would once again inherit the earth."[36]

The ARS saw *Silent Spring* as a direct assault on its policies. ARS staff did not leave behind an ideological mission statement, but memoranda, letters, reports, and studies reveal that the agency demonstrated a disturbing compulsion to advocate and defend pesticides. It had close ties with farm and industry organizations, land grant universities, experiment stations, the federal extension service, and many farmers. The ARS possessed enormous power, for its label approval function licensed pesticide formulations. It garnered enormous power in its multiple roles as clearinghouse, coordinator, regulator, and research center. To have their way, ARS bureaucrats bullied, plotted, lied, and misled. A culture emerged within the service that justified pesticides at all costs, and staffers bent research, reports, and testimony to serve this mission.

ARS administrator Byron T. Shaw's analysis of Carson's *New Yorker* articles, which was sent to Secretary of Agriculture Orville Freeman, epitomized the ARS mindset. While admitting that Carson's work "should not be dismissed lightly," Shaw faulted her "repeated use of fragmentary data or isolated cases to support broad generalizations." Her biased approach, he warned, "serves only to frighten people unnecessarily" while threatening the food supply. He attempted to undermine her balance-of-nature argument by citing agricultural production statistics that claimed 8 percent of the population fed 186 million people, while a century earlier farmers had composed 60 percent of the United States' total population of 31 million. "As a practical matter," he argued, ignoring the USDA's role in pushing people off the land, "we would have no farms in the United States if we depended upon nature's unaided balance." Shaw ignored the vanishing farm population set in motion by USDA policies and by science and technology, as did most USDA bureaucrats. That fewer and fewer farmers could feed more and more people meant that the farm population would continue to decline, but to reduce chemicals to increase the number of farmers was heretical. Shaw also championed chemical intrusion as necessary to protect consumers from rot and filth. "The apples we bite without fear of worms, the oranges we squeeze without fear of maggots, and the countless other foods we eat every day are possible because modern man has chosen to assist nature," he asserted.[37] Before USDA-guided chemical intervention, Shaw implied, nature had failed miserably.

Rodney E. Leonard, assistant to the secretary, warned Freeman that Shaw's analysis was "a typical bureaucrat response which never quite joins the issue and further, doesn't recognize the emotional impact which the Carson articles have on people." Shaw's statements, he warned, would create "a strong adverse public reaction." Leonard understood that Carson was not, as Shaw portrayed her, blindly antichemical. "She is arguing against the ignorant use of deadly compounds," Leonard explained, "and for more adequate research as to the immediate and the cumulative effects of these chemicals." Leonard thought that "she makes a lot of sense," although "she overstates her case in a number of instances." He suggested to Freeman that the USDA would profit from taking *Silent Spring* seriously. Still, when ARS staffers attended a meeting of the President's Pesticide Task Force on August 23, 1962, with representatives from the Public Health

Service, the FDA, the FWS, and other agencies, they were surprised and dismayed to discover that other agency staffers gave "the impression they agreed with and believed most of the Rachel Carson articles."[38] Carson's book frustrated ARS bureaucrats in part because it was a sophisticated synthesis of existing scholarship, supplemented by informants in other agencies rather than by original scientific research.

If Freeman's response to William C. Parrish of Memphis was any indication, he was no disciple of Rachel Carson. Parrish had just read *Silent Spring* and asked Freeman if there were plans to investigate pesticide threats. Freeman granted that *Silent Spring* was "a well-written compilation of incidents and opinions," but he argued that Carson had taken a complex subject and had arrived at "unwarranted conclusions." He explained to Parrish the department's responsibilities in registering pesticides and the Department of Health, Education, and Welfare's role in enforcing food purity. Freeman claimed that chemicals "made it possible for the American people to enjoy the greatest abundance of wholesome food ever known."[39] Freeman, like others in the USDA, stressed the role of chemicals in production but not, of course, in overproduction.

In the spring of 1963, Clarence Cottam declared that the country stood at a major turning point with regard to chemicals. "We've got to decide whether we follow the path of sanity, maturity and judgment or go down the road labeled 'we haven't died yet so it's all right,'" he proclaimed. In an interview he stressed the need for better control of chemicals and more research on toxicity. At the annual meeting of the National Wildlife Federation, Cottam warned against the quarter-million dollar campaign launched by chemical companies to discredit *Silent Spring.* He named no names but suggested that some scientists might be involved in payola. "Who is getting the gravy?" he queried pointedly.[40]

In its defense of pesticides, the USDA argued that pesticides were necessary and effective while also asserting that pests remained a major threat to food production. From 1951 through 1960, an ARS study reported, pest-caused losses to crops, forests, livestock, and poultry totaled $14.3 million; pest-caused storage losses totaled $2.3 million. These numbers can be read as a justification for more pesticides, an admission of the ARS's failure to control pests, or a statement on overproduction during the 1950s. Indeed, USDA policies were a jumble of contradictions. It funded insect-control projects, research that encouraged higher agricultural production, programs such as the Soil Bank and acreage allotment cuts (which reduced production), and costly storage of surplus commodities. In admitting insect attacks on stored commodities that resulted from government-sponsored research, the ARS report, no doubt unconsciously, supported the argument that pesticides were not necessary to produce an adequate food supply. Although informed critics such as Clarence Cottam dissected USDA logic and called its hand on policy and operational contradictions, the department's size, power, and ruthlessness protected it from effective public opposition and endeared it to the agribusiness community and to congressional backers.[41]

4
EXPERT TESTIMONY

Contamination of various kinds has now invaded all of the physical environment that supports us—water, soil, air, and vegetation. It has even penetrated that internal environment within the bodies of animals and of men.

Rachel Carson, May 16, 1963

Silent Spring sent tremors through the USDA bureaucracy, which became more defensive and dodgy as public opinion turned against pesticide use. In response to growing concerns over pesticides, President John F. Kennedy appointed Jerome B. Wiesner, a professor at the Massachusetts Institute of Technology, to head the President's Science Advisory Committee; the president asked Wiesner and the other scientists and engineers on the committee to study the impact of pesticides. When in February 1963 Wiesner circulated a draft report on pesticides, the ARS's T. C. Byerly sniveled that it "singles out agriculture as a miscreant, the pesticide industry as culpable, and in effect, leaves the health agencies and wildlife agencies blameless." ARS administrator Byron T. Shaw anxiously recited pesticides' virtues. "Pesticides, which rank among the significant scientific developments of the century, have pushed us far along the road of human progress," he argued. Without pesticides, Shaw claimed, some 70 percent of the country's important crops could not be grown. He feared that releasing the report would "profoundly damage U.S. agriculture." Offering a benign, albeit irrelevant, analogy, Shaw asserted that "vital statistics show that accidental fatalities caused by all pesticides annually will about equal those caused by aspirin alone, while those caused by sleeping tablets are more than two times as many." After denouncing the draft report for devoting only four pages to the "benefits" of pesticides and fourteen to their "hazards" and fretting that it might provoke European trade barriers, Secretary of Agriculture Orville Freeman also resorted to dubious analogies. "The hazards inherent in the use of gas, electricity and certain patent medicines, for example, certainly outweigh the dangers from the use of pesticides," he proclaimed. At the time Freeman made his statement, the ARS had no central file that accumulated and analyzed national poisoning data. Given that as of June 1962, the ARS had approved some 54,000 formulations containing 500 chemical compounds, Freeman relied upon faith rather than information in making this claim.[1]

From February to May 1963, the USDA, the Department of Health, Education, and Welfare (HEW), and the Department of the Interior battled over the report's final wording. USDA staffers warned that a report unsympathetic to pesticides could undermine food production, and HEW worried that the public would interpret the report as a statement that food was unsafe. As this debate swirled through the bureaucracy, the pesticide issue reached prime-time television when CBS broadcast *The Silent Spring of Rachel Carson*. The committee released its report, "Use of Pesticides," on May 15, 1963, and the next day Wiesner appeared before Senator Abraham Ribicoff's subcommittee on Reorganization and International Organization. The hearings became a forum for debate between those who defended and those who criticized pesticides. Although the committee's report had been modified to address some ARS objections, it nevertheless called for a "more judicious use of pesticides or alternate methods of pest control." Few systematic studies of occupational exposure to pesticides had been done, the report admitted, but it did cite, without attribution, Wayland J. Hayes Jr.'s prisoner study, which showed no ill effects from taking a daily dose of 35 mg of DDT. Each year, the report continued, roughly 150 people died from pesticide poisoning, half of them children. Wildlife losses occurred both because of "carelessness or nondirected use" and because of "programs carried out exactly as planned." It mentioned bird losses in areas sprayed with DDT to control Dutch elm disease and wildlife losses in areas sprayed for fire ant control. The report discussed pesticide approval, tolerances, and terminology, along with the relative toxicity of chlorinated hydrocarbons and organophosphates. It called for HEW to sponsor data collection and studies of occupational exposure and to cooperate with other federal agencies. The FDA needed to launch a major effort to complete its "review of residue tolerances." In order to reduce chemicals in the environment, the report suggested that research be undertaken on more selective, nonpersistent, and nonchemical control methods. Finally, it called for reliable studies relating to human health and threats to wildlife.[2]

Secretary Freeman testified after Wiesner. He parroted the tired USDA claim that the department guaranteed safe food that fed a hungry world.

Chemicals, he stressed, were the key element in maintaining farm productivity. If there were dangers in using pesticides, Freeman suggested, "there may be greater dangers in not using them." After reviewing the bureaucratic route to pesticide registration, he recommended ending the "under protest" category, which allowed a manufacturer to market a product until the USDA investigated its toxicity and took legal action. Freeman reviewed the fire ant control program and admitted that ARS initially used two pounds of heptachlor per acre, which "effectively killed fire ants but . . . also killed some desirable fish and wildlife." Reducing the amount of heptachlor, he suggested, killed fire ants "with virtually no hazard to wildlife." The secretary excitedly reported a 1961 "breakthrough" in fire ant control, when mirex, another chlorinated hydrocarbon, became available. "It has no harmful effects on people, domestic animals, fish, wildlife, or even bees, and it leaves no residue in milk, meat, or crops," he avowed. Nearly every claim in that hopeful statement dissolved under research scrutiny.[3]

After the hearing, the press asked Freeman for the names of the products marketed under protest, but he refused to reveal them. "I believe the Department's action is utterly indefensible," Senator Ribicoff railed at his subcommittee's next hearing on June 4, 1963. He declared that "it is a mockery of regulation for the Department of Agriculture to find a product unsafe and then refuse to tell the public the name of the product." He demanded that the USDA provide a complete list of products sold under protest registration "by the end of the day." The Department of Agriculture complied.[4]

Ribicoff, Kennedy's former HEW secretary, then welcomed Rachel Carson to the hearing. "The contamination of the environment with harmful substances is one of the major problems of modern life," Carson began, and she continued with an eloquent warning about the fate of the earth and its creatures. "Contamination of various kinds has now invaded all of the physical environment that supports us—water, soil, air, and vegetation," she cautioned. "It has even penetrated that internal environment within the bodies of animals and of men." Mankind had produced a discouraging record, she judged, "for that history is for the most part that of the blind or shortsighted despoiling of the soil, forests, waters, and all the rest of the earth's resources." The effect of pesticides within human bodies had been ignored, she contended, and scientists had not investigated how drugs interacted with body-stored pesticides. She chronicled how pesticide drift had become a universal problem and observed that DDT had been found in remote areas that were far from any human activity. Aerial spraying, she explained, launched small crystals into the air, "the components of what we know as 'drift'—the phenomenon that plagues every householder who receives contaminating spray from his neighbor across the street, or from his Government's spray planes several miles away." She testified that 2,4-D had drifted fifteen to twenty miles, causing vegetation damage. Pesticides were also washed off the land by rainwater and carried into streams and ultimately to the oceans. Animals transported pesticides in "a natural food chain, usually becoming concentrated as it goes." Aerial pesticide application "should be brought under strict control," she suggested, arguing that persistent pesticides should be reduced and eventually phased out. Speaking "not as a lawyer but as a biologist and as a human being," Carson defended "the right of the citizen to be secure in his own home against the intrusion of poisons applied by other persons." She suggested that the sale of pesticides be restricted, since many users had no understanding of their hazards. Then she focused on her larger target. "I should like to see the registration of chemicals made a function of all agencies concerned rather than of the Department of Agriculture alone," she proposed. The proliferation of chemical compounds, she concluded, "is dictated by the facts of competition within the industry rather than by actual need."[5]

After Carson's statement, Ribicoff asked that she clarify her position on chemicals, for the public saw a sharp conflict between her views and those of agricultural chemical manufacturers. Carson admitted that much of the criticism of *Silent Spring* had "been placed on an all-or-none basis, which is not correct." After agreeing that chemicals had produced some benefits, she stated, "I think that we have had our eyes too exclusively on the benefits, and we have failed to recognize that there are also many side effects which must be taken into consideration."[6]

Three weeks after Carson testified, Ribicoff's subcommittee heard representatives of the chemical industry respond to the President's Science Advisory Committee report. On June 25, 1963, Parke Brinkley, a former Virginia agriculture commissioner and head of the National Agricultural Chemicals Association, testified for the chemical industry. He spoke for several different organizations: the Manufacturing Chemists' Association, which was founded in 1872 and included 199 companies; the Chemical Specialities Manufacturers Association, founded in 1914 with 500 member companies; and the National Agricultural Chemicals Association, founded in 1933 with 112 companies. Brinkley boasted that these companies produced 90 percent of the United States' basic chemicals and 85 percent of its formulated pest control chemicals; they dominated a $300 million industry. These powerful associations had gone to war against Rachel Carson and, through expert witnesses, had participated in the *Lawler* case. Brinkley correctly observed that the impact of mechanization on rural life was more apparent than "chemicals which protect crops from insect, weed, and disease damage." He linked chemicals to the explosion of suburban development in the post–World War II United States. "The control of these and other insects also permits the patio-type living which has become so popular in our suburban areas," he boasted. Chemicals protected houses from termites, he continued, and saved people from disease. He described a number of control programs that supposedly had contributed to Americans' health and happiness. Brinkley favorably quoted Wayland J. Hayes Jr.'s claim that "it has been impossible to confirm the allegation that insecticides, when properly used, are the cause of any disease of man or animals." Arguing that there was already too much government intrusion in the chemical industry, he disagreed with those who advocated shelving persistent pesticides. In a point-by-point rebuttal, Brinkley disagreed with the Wiesner report, claiming that "the general tenor of the report is that controls are inadequate and need to be strengthened." After asking for more research into Carson's charges, Brinkley yielded the floor to Mitchell Zavon.[7]

When asked who he would recommend for expert testimony before the committee, Zavon mentioned, among others, Wayland J. Hayes Jr. and Griffith Quinby, two of the physicians who participated or were cited in the *Lawler* case. Senator Ribicoff established that Zavon reported cases of human pesticide poisonings to Hayes; yet even as he did so, Zavon did not recommend that Hayes's office tabulate all such reports, claiming that it would "clutter up our files." Zavon advocated long-term, not isolated, studies. Ribicoff scoffed at Zavon's reluctance to track all human poisonings, asking, "How is even Dr. Hayes of the Communicable Disease Center and his staff going to get this information if you people have it and other manufacturers have it, and you don't give this information to Dr. Hayes?" In a sentence that harkened back to the *Lawler* case, Zavon stated, "The diagnosis of pesticide intoxication requires a history of exposure or a reasonable possibility that such exposure could have occurred; without such a history, it is safer to look elsewhere for the diagnosis." Pesticides were safe when used as directed, he insisted, but problems had "arisen from gross misuse or from willful misuse, such as suicide attempts." He denied that pesticides produced long-term health effects, and he also argued that the American Medical Association did not need to conduct more studies on the subject. Zavon set a high bar for proof that pesticides were in any way linked to illness, dismissing outright thousands of reported poisoning cases. Brinkley then resumed his testimony, stressing that his mission was to "renew the confidence of the American public in research scientists."[8]

Wayland Jackson Hayes Jr. identified himself to the committee on July 17, 1963, as the chief of the Toxicology Section at the Communicable Disease Center (CDC) in Atlanta. Born in 1917 in Charlottesville, Virginia, he received his Ph.D. from the University of Wisconsin in 1942 and his M.D. from the University of Virginia in 1946. After an internship at the Public Health Service (PHS) hospital on Staten Island, New York, he became chief of the PHS toxicology section in Savannah, Georgia, in 1947, where he worked on pesticides and human health. In 1960 he moved with the toxicology section to the Atlanta CDC. He was a prolific researcher and writer. Over the years he became one of the most visible and vocal defenders of DDT. In 1968 he left the PHS to teach at the Vanderbilt University School of Medicine.[9]

The post–World War II proliferation of synthetic chemicals, Hayes testified, had led to "no increase in the rate of fatal poisoning." Aspirin, he argued, was less toxic than parathion but "is a more important cause of death." There was no evidence that pesticides "old or new," he continued, "are a cause of any disease except poisoning." He cited his 1956 study that men could eat DDT at "a level approximately 200 times greater than that in the ordinary diet without showing any detectable clinical effect." Ribicoff pressed Hayes on his blanket statement that pesticides were benign, whereupon Hayes equivocated, as did so many chemical advocates, by arguing that more research was needed.[10]

On July 17, Ribicoff's subcommittee heard from physicians who researched and treated diseases triggered by chemicals. One of the witnesses was Dr. Malcolm M. Hargraves, who had graduated from the Ohio State Medical School and had gone to the Mayo Clinic in 1935, specializing in hematology. The chemical industry criticized Hargraves for his earlier congressional testimony, and his public stance on pesticide dangers did not sit well with some of his colleagues. Hargraves and other physicians and scientists who found chemical causes of disease directly contradicted

Mitchell Zavon, Wayland Hayes Jr., and Griffith Quinby, who saw no harm from pesticides if used as directed. Hargraves had shared his research with Rachel Carson. "On the basis of such observation and ecologic consideration as I have mentioned," he began, "I believe that the leukemias and lymphomas, as well as many other diseases, are produced in the genetically susceptible person as a reaction to certain environmental situations involving agents such as the hydrocarbons, infections, actinic rays, irradiation, and others." He ranked benzene as "the agent most responsible for the production of blood dyscrasias," but other hydrocarbons were suspect as well. People's susceptibility to chemically induced illness varied greatly, he explained. "It is the chronic, insidious, sublethal, intermittent, and recurrent exposure that is significant," he stressed. Most people adjusted successfully to a pesticide load, but he offered several case histories of those who did not. He began with a farmwife who was exposed to a lindane vaporizer and over several years became anemic. By the time Hargraves saw her, she had received a dozen blood transfusions, and a bone marrow examination "showed a complete red cell dyscrasia." Her husband tossed out the vaporizer and washed off the walls of her room, and she recovered. Hargraves discussed other cases involving varnish remover, cleaning compounds, moth spray, hair dye, paint, and lindane. His patients, he explained, "give this type of history of exposure to gasoline fumes, to tractor fumes, to the use of insecticides, and very often we can correlate an exacerbation of their disease with a reexposure. It is very common for them to occur after planting time or after having sprayed for weeds." He explained that his patients invariably were exposed to chemicals, saying, "Or, to put it another way, physicians, engineers, executives, teachers and housewives who are not exposed to these agents simply do not seem to find their way to my office afflicted with mesenchymal dyscrasia."[11]

Dr. Joseph H. Holmes, head of the division of renal diseases at the University of Colorado, then testified about his extensive experience working with organophosphates. Holmes held a M.D. degree from Western University and a doctorate in medical science from Columbia University. He began studying organophosphates in 1950, he testified, when working on a contract for the Chemical Warfare Service dealing with a Rocky Mountain Arsenal project on nerve gas and insecticides. His research between 1953 and 1961 showed that many people in the area absorbed organophosphates, especially those who serviced the airplanes used by crop dusters. Around Denver he also found organophosphate exposure in people who worked in processing plants, among florists and greenhouse workers, around homes, and among children. Fortunately, he testified, no one had died, but seventy cases "required emergency hospital treatment." Holmes insisted that pesticides needed better labels. When he had pointed out dangers to users, he said, "they were not aware of the toxicity because the label is in fine print and at the end of the using instructions rather than at the beginning." The fact that pesticides were sold in supermarkets "along with groceries" also disturbed him. Although there was no reliable data on long-term exposure to pesticides, Holmes had traced test performance and found that people exposed to pesticides regularly took longer to perform tasks and made more errors than others. Duster pilots, he assumed, sometimes crashed because they were overcome by organophosphates. His psychological testing revealed worker impairment and a "don't give a damn" attitude. One worker "said that he couldn't adjust as well to his work and his family, and he complained particularly of inability to concentrate, forgetfulness, and having a continual argument with the foreman and his family."[12]

On July 18 the National Cancer Institute's Dr. W. C. Hueper testified before Ribicoff's subcommittee, painting a dire picture of pesticide risks. Hueper explained that many of the synthetic chemicals developed since World War II "display a considerable toxicity to the liver, alimentary tract, nervous system, and skin of man." Developed as chemical warfare agents, he explained, organophosphates were effective in "exceptionally small amounts, and recently have been suspected of affecting the function and integrity of the central nervous system, following repeated exposures to doses previously considered as harmless or causing only transitory effects." Even chlorinated hydrocarbons, he continued, "are exquisite liver and kidney poisons to man and animals." Many pesticides target the same organs—the liver, kidney, and brain—he warned, "and simultaneous and sequential exposure to a combination of such chemicals may result in a summation, aggravation, accentuation, and potentiation of the toxic effects of the individual chemicals." Hueper was distressed that such powerful chemicals were "freely available on the open market to the general public." He recited a chilling list of "cancer producing" chemicals and offered a set of suggestions about how to curb pesticide dangers. Animal studies, he pointed out, raised serious questions about the long-term effects of pesticides. Hueper branded an Esso scientist's claim that there would be no "epidemiclike occurrence of cancers" in the next ten to thirty years as "whistling in the dark." Such scientists "arrogantly" held out for absolute proof before agreeing to any governmental intrusion. This, Hueper concluded, was an argument that "private parties essentially interested in monetary profit obviously would receive what amounts for all practical purposes to a license to harm and kill fellow citizens, if these parties should deem such a procedure in the interest of economic progress for themselves and incidentally and possibly also for the Nation." He vehemently disagreed: "I do not believe that such a social philosophy is in the public interest."[13]

In this battle of expert witnesses, one could find security or horror in pesticide use. As Hayes and Zavon argued, attaching blame for a health problem to a pesticide was often extremely difficult. Hayes's research supported the position that DDT and most chlorinated hydrocarbons had little if any observable effect on human health. Animal studies were more problematical. Other scientists feared that long-term pesticide effects would reveal serious health implications. In the 1950s and 1960s, the debate swirled around the work of relatively few scientists and physicians. And as PHS researchers, both Wayland J. Hayes Jr. and Griffith Quinby were in sensitive yet powerful positions to affect policy.

In the summer of 1963, for example, Quinby accompanied a USDA employee to Toppenish, Washington, to investigate a pesticide incident that had hospitalized thirty-eight people and killed two cows. Washington senator Warren G. Magnuson demanded of Secretary of Agriculture Orville Freeman a full report. "While the Department of Agriculture may find no obligation to protect our fish and wildlife," Magnuson scolded, "it occurs to me that insecticides which kill domestic livestock and hospitalize humans must give the Department grave concern, and, as a corollary, insecticides that kill domestic animals may also be harmful to wildlife." Freeman explained that the incident had resulted from spraying TEPP from an aircraft, "a private undertaking." Quinby and the USDA agreed that "this was a rare occurrence due to freak weather conditions, that it was unavoidable and unpredictable." The TEPP was properly labeled, Freeman added.[14] Ultimately, Quinby's report raised more questions than it answered.

By August 1963 the USDA's George Barnes concluded that the department had weathered the storm caused by *Silent Spring*. "In general," he confided to Secretary Freeman, "I think we can say that the pesticides matter, which once threatened to embarrass the Department, is in hand." He credited both Freeman's Senate testimony "and the calm but effective way ARS has responded in plugging loopholes, revising procedures, etc."

George A. Barnes, July 19, 1961.
16-BN-14038, National Archives and Records Administration (NARA).

For the USDA, pesticide dangers presented a public relations problem, but cosmetic adjustments in the ARS hardly addressed the larger problem posed by toxic chemicals. Indeed, in April 1964, Freeman met with ARS staff to discuss the increasing sophistication of pesticide detection, which revealed previously unknown residues. In such cases, he surmised, "we are actually probably acting illegally as well as being in the impossible position of endangering human life by not revoking registrations." Freeman wanted to act quickly and "not continue as defensive as we are now and will be each time an incident comes to light and some demagogues around town have an opportunity to exploit it."[15]

Despite the impact of *Silent Spring,* pesticides were so imbedded in agricultural practice that application schedules were sometimes only theoretically tied to actual pests. In the summer of 1964, Nyle C. Brady observed that in Texas pesticide applications were based on "the predicted appearance of harmful insects" and that the "fixed application schedule does not take into account whether insects are actually present." Brady had a good reason for his concern, because the USDA had recently announced a policy of minimizing pesticide applications. "There is a presumption," he reasoned, "that insecticides might be used unnecessarily in the absence of insects." He appointed a task force to sort out the problem.[16]

Government scientists seemed to prefer hypothetical cases to unique and perhaps important information that materialized from the public. On October 12, 1964, for example, Dr. Kenneth M. Lynch, who raised pure-blood wild turkeys, wrote to the PHS about his observations. Since 1957, he began, Summerville, South Carolina, had been fogged each summer with DDT or malathion to kill mosquitos. Over the past several years, some of his turkeys "have more or less abruptly developed progressive paralysis, from apparent good health to death in two or three days." At the same time, egg hatchability "has been reduced from an expected 75% or so to less than 10%." Lynch had discussed these symptoms with a professor of chemistry, and the pair wanted to apply for a grant for further study. The PHS offered no support, but Lynch appealed again in November. He insisted that his turkeys offered the opportunity to study the impact of pesticides on the community's wildlife, for his turkey pen was located in the middle of town and was isolated from other chemicals. "The evidence already secured of the apparent effects of insecticide fogging, directly on the birds and upon their reproductivity, seems to require that their tissues and their eggs should be chemically examined by the best techniques," he argued. Nonetheless, Dr. Lynch got no encouragement from Washington.[17]

Clarence Cottam, meanwhile, had uncovered more self-serving and erroneous ARS statements. In May 1965, he complained to Secretary Freeman regarding an address by Dr. Clarence H. Hoffman, the assistant director of the ARS's Insects Affecting Man and Animals division. Hoffman denied that DDT had any negative health effects on humans and denounced Rachel Carson's concerns about pollution. Cottam cited several wildlife studies to refute Hoffman's claims. Hoffman knew better than he spoke, Cottam charged, which "leaves me with the inescapable conclusion that his comments are indicative of extreme bias and an unyielding rigidity within Agriculture and particularly within ARS." Cottam stopped short of branding Hoffman untruthful or unethical, instead tying his statements to ARS ideology. "It seems to me it can hardly be interpreted as anything other than a case of bureaucracy being blinded by its own past mistakes," he judged.[18]

Nyle C. Brady replied to Cottam that press coverage did not fully convey Hoffman's concerns about DDT. When questioned about DDT hazards to humans, for example, Hoffman cited Dr. Wayland J. Hayes Jr.'s well-known experiments with prisoner volunteers. Cottam replied that he was acquainted with Hayes's studies, including the experiment with prisoners. He recalled that volunteers could withdraw from the experiment at any time and that "there were a fair number who withdrew when they became a bit ill," which undermined the study by eliminating negative evidence. Hayes had been under stress for his stand on pesticides, Cottam suggested, adding, "Perhaps he is like many human beings who when subjected to criticism become more and more dogmatic in maintaining their initial stand." Hoffman's one-sided presentation, Cottam concluded, "is a little less than being intellectually honest and objective."[19] Yet for most people, the press report of Hoffman's talk gave the official lie to DDT's safety. Not only Hoffman but the ARS knew better, but the bureaucratic imperative to protect pesticides led the division into territory alien to honesty.

Ribicoff's hearings, in addition to airing opposing views of pesticide dangers, suggested the size of the opposing armies. There were over eight hundred chemical companies opposed to Rachel Carson and her environmental allies. Strangely, chemical companies reacted hysterically to *Silent Spring,* predicting hunger and doom if farmers were deprived of pesticides. In Mississippi, the Delta Council and Farm Bureau, which both had been obsessed with the *Lawler* case and the possibility of subsequent challenges to the aerial application of pesticides, suddenly saw an even larger threat from *Silent Spring.* B. F. Smith, executive vice president of the Delta Council, sent members a letter appraising them of Carson's arguments and a copy of "The Desolate Year," a *Monsanto Magazine* fantasy of insect domination. "The bugs were everywhere. Unseen. Unheard. Unbelievably universal," it announced. Without chemicals, insects destroyed crops, infested domestic animals, pestered humans, and ruined the economy. Monsanto's story of insect domination was part of an effort to counter Carson's claims. Congressman Jamie Whitten charged that *Silent Spring* "frightened our public and the people haven't gotten over it yet." He lamented that "hundreds of millions of dollars" had been spent in "indemnities and in research trying to quiet a public feeling," even though there was no indication that pesticides harmed people when used properly.[20] The aim, obviously, was to create panic and have farmers rush into the arms of the chemical companies.

Whitten's 1967 book, *That We May Live,* was both a puerile effort to discredit *Silent Spring* and an unyielding defense of pesticides. The book had been conceived and subsidized by chemical companies. The *Washington Post* learned that Samuel Bledsoe, who worked for a public relations firm, had suggested that Whitten write a book countering Rachel Carson. Bledsoe's firm represented Velsicol, which at the time was busily denying any role in the Mississippi River fish kills. Parke Brinkley, president of the National Agricultural Chemicals Association, accompanied Whitten and Bledsoe to a meeting with a potential publisher, M. B. Schnapper of the Public Affairs Press in Washington. They indicated that "a large number of copies would be bought by interested industrial parties," Schnapper recalled. He considered the book "substantively weak and was poorly written" and also felt "squeamish about what was going on in terms of industrial sponsorship of the book." D. Van Nostrand eventually published *That We May Live* but demanded a subsidy from Velsicol, Shell, and CIBA-Geigy. One chemical company executive estimated that CIBA-Geigy bought five hundred copies of the book, while Velsicol bought two thousand. None of the parties were pleased with the project. "It left something to be desired," a CIBA-Geigy spokesperson admitted. "It was less than scholarly." The backers possibly reasoned that getting *That We May Live* published, while not adding to scientific discourse, might humor Congressman Whitten and help their cause.[21]

Silent Spring legitimized debate over pesticide dangers to wildlife, human health, and, on a larger scale, the earth's environment. Prior to Carson's work, pesticide advocates had often portrayed environmentalists as befuddled alarmists who had no contact with real-world concerns. *Silent Spring* provoked President John F. Kennedy to appoint Jerome B. Wiesner to head the President's Science Advisory Committee, and the ensuing congressional hearings on Wiesner's report gave both environmentalists and physicians concerned with pesticide-induced sicknesses a chance to make their voices heard. Still, while Carson's book resonated with growing fears about pesticides and spurred the modern environmental movement, it did not halt the growing addiction to agricultural chemicals. Adept at persuading the public that their products were benign, the chemical industry used advertising and expert testimony to insist that pesticides offered no danger. The public was pulled between advertisements brimming with promises and assurances on the one hand and cautious physicians and authors who warned of pesticide dangers on the other. The ARS was assigned to umpire the debate and assure public protection, yet it had become simply the handmaiden of the chemical companies.

5
MISSISSIPPI RIVER FISH KILLS

We are not yet ready for anyone to make a judgment that fish and wildlife must be sacrificed on the alter of agricultural overproduction.

Thomas L. Kimball

In the early 1960s, the press reported massive and disturbing fish kills in the lower Mississippi River, the stretch from the northern border of Arkansas to the Gulf of Mexico. The Public Health Service (PHS) counted fourteen major kills throughout the country in 1961, ten in 1962, and thirteen in 1963. Minor fish kills were more easily solved. A Kentucky report, for example, indicated that "outlaws" used rotenone to stun fish, continuing, "This type of deliberate poisoning of fish occurs about 4 or 5 times each year in Kentucky." Fish kills, the PHS discovered, were linked to chlordane, 2,4-D, 2,4,5-T, toxaphene, copper sulfate, heptachlor, dieldrin, parathion, rotenone, and endrin. The Mississippi River kills, however, were baffling. Dying fish rose to the surface bleeding from the mouth and fins. Scientists searched in vain for clues, but water quality, temperature, and food supply appeared satisfactory. In frustration, scientists declared "abdominal dropsy" the villain. Another major kill in the lower Mississippi River in the fall of 1963 provoked civic and scientific concern. Taking advantage of the latest detective and analytical devices, PHS scientist Donald I. Mount discovered deadly amounts of endrin in the water and mud, as well as in the fish. On March 17, 1964, the PHS released Mount's findings.[1]

At almost the same time that the PHS released its report, the press revealed another massive kill in the lower Mississippi and Atchafalaya rivers. On March 22, 1964, the *New York Times* began a series of stories that dramatized the problem. In Franklin, Louisiana, reporter Harry McHugh marveled at the power of the unknown poison. "Any time it brings up to the surface big blue catfish weighing 35, 50, 75 pounds apiece, it is quite potent," he wrote. He had seen a dead gar that weighed 150 pounds. Trappers and sportsmen spotted other disturbing signs that a deadly poison was coursing through the environment. Crows and white cranes had become scarce along the Atchafalaya River, and there were reports of dying ducks falling from the sky. Dr. James R. Strain, president of the Louisiana Board of Health, reported that shrimp had also been killed by chemicals; he suspected endrin, dieldrin, heptachlor, and DDT. Strain worried about the potential health effects on people who ate shrimp and oysters. The Board of Health asked for a federal investigation into the crisis.[2]

Senator Abraham A. Ribicoff immediately announced hearings before his subcommittee of the Committee on Government Operations. The senator had sparred with Agricultural Research Service (ARS) bureaucrats for years over inadequate pesticide labeling, and he assumed that pesticide residues washed from farmland killed the fish. Ribicoff was weary of the ARS cliche that no harm would come from pesticides if used according to label instructions. The fish kills prompted the *Washington Post* to editorialize that the crisis vindicated Rachel Carson's warnings. The paper observed that the British government had recently banned aldrin and endrin, and it spoke out in support of Senator Ribicoff's hearings. "His subcommittee," the editorial concluded, "ought to make a realistic appraisal of the safeguards necessary to end robin-counting as an unpleasant new rite of spring."[3]

The USDA realized that if endrin runoff from farms was responsible for the fish kills, their approval and labeling policy could come under scrutiny. On April 1, two ARS scientists, along with representatives from Shell and Velsicol, attended a meeting at the PHS's Taft Sanitary Engineering Center in Cincinnati. The agenda restricted discussion to "analytical methods and technical procedures used in the fish kill studies." The scientists reported that "there is strong evidence that endrin was responsible" for the kills, continuing, "It was shown that endrin was present in the river water and in the blood and other tissues of dead and dying catfish in amounts sufficient to account for their death." The chemists did not speculate on how the pesticides got into the river. Nyle C. Brady, director of science and education at the ARS, forwarded a summary of the Cincinnati meeting to Secretary of Agriculture Orville Freeman. Despite Donald Mount's impressive research, Brady branded the report inconclusive. "Although there is a strong indication that insecticides may have been responsible for the fish kill," he quibbled, "there is as yet no conclusive proof that this is the case." He stressed that the autumn 1963 kill came at a time of low water, "when one would expect runoff and drainage from agricultural lands to be at a minimum."[4]

On April 3, Brady crafted a supportive memorandum to Freeman. Brady complained that the PHS had not notified the ARS of its study or the results until March 13, four days before its release. He conceded that "certain pesticides, under certain conditions and in certain amounts, can and do kill fish," but, he fretted, the study was "by no means sufficiently conclusive to warrant the reports which have been given widespread currency by the press." Already worried about potential questions at the upcoming Ribicoff hearing, Brady stressed that the report did not locate the source of endrin or other chemicals. Attempting to shift the cause of the fish kills from agricultural use to manufacturing plants, he suggested that the poison might come from the "54 industrial plants along the Mississippi River in Tennessee, Arkansas, Mississippi, and Louisiana which manufacture pesticides, fertilizers, and pesticide-fertilizers." The ARS secured revisions in the original PHS draft report "on the ground that it suggested premature conclusions," and the PHS accepted modifications "that unwarranted emphasis would not be placed on pesticides as the agent responsible for the fish kill." Brady reported that the ARS would conduct public hearings in Washington and Memphis in mid-April. If pesticides were responsible for the kills, Brady assured Freeman, the ARS would take the necessary actions to end endrin registration. He stressed that the PHS's "unilateral approach" had embarrassed the USDA and "created premature public concern." Brady's petulance about "the lack of coordination among executive agencies interested in pesticides and their use" was a turnabout for the USDA, which only stingily shared any information with other agencies. On April 5, the *New York Times* ran a story alleging that additional PHS laboratory tests showed that endrin was indeed the culprit in the fish kills, and that Shell and Velsicol were endrin's chief producers.[5]

On April 7, 1964, James M. Quigley, assistant secretary of the Department of Health, Education, and Welfare (HEW), led off the hearings before Senator Abraham Ribicoff's subcommittee. Quigley reviewed the extent of fish kills from 1960 to the fall of 1963. "Catfish, drum, buffalo, and shad were observed swimming near the surface of the river and dying," he testified. "In the brackish waters near the Gulf of Mexico, whole schools of menhaden and numerous mullet, sea trout, and marine catfish died." He estimated that some 10 million fish had perished from poison in the past four years. HEW scientists were at first baffled. "Fish might die upstream from Baton Rouge one week, on the delta below New Orleans another week, and somewhere in between these points at a later date," the PHS report observed. After ruling out natural causes, scientists suspected pesticides. "Extracts of the mud taken from the area of the fish kills, when placed in aquaria containing healthy fish, resulted in death," it found. "Extracts of the tissues of dying fish, when dissolved in the water of the aquaria, also resulted in the death of healthy fish." Relying upon newly developed gas chromotography technology that measured to parts per billion, scientists discovered residues of endrin, dieldrin, DDT, DDE, and other synthetic organic chemicals in the fish tissue. "Endrin," Quigley reported, "was consistently found in all dead fish tissue extracts examined." When Senator Ribicoff asked about endrin's toxicity, Donald Mount gave an example: "I understand, sir, that three drops of pure endrin in the Potomac River would be one part per billion and that would kill fish in the water." The PHS report theorized that fish stored endrin in their fatty tissue and, during the winter when the fish were not feeding, these tissues "are reabsorbed and the accumulations of endrin may be released into the blood at lethal levels." Catfish concentrated endrin at levels a thousand to ten thousand times greater than found in river water. James M. Hundley from the PHS expressed concern that "there is a health hazard in this situation in the Mississippi, both with respect to drinking water, and with respect to food that may be coming from the Mississippi River." He was not sure how great the hazard might be. The PHS report on fish kills observed that "dead and dying catfish" were "sold for human consumption." Donald McKernan, director of the Department of Interior's Bureau of Commercial Fisheries, estimated that 80 million pounds of fish were taken from the Mississippi River each year for commercial purposes. The fish kills would surely take a toll on fishermen's livelihoods.[6]

Environmentalists, sportsmen, and public health officials were less concerned about localized kills that could be traced to a specific chemical spill or accident than about the massive and mysterious kills along the Mississippi River. Something invisible and deadly had infected the water moving toward the Gulf of Mexico. "The particularly alarming aspect of the findings on the lower Mississippi," the PHS report concluded, "is that these kills are much more widespread; their long and recurring nature suggests a continuing contamination of the environment." While the PHS report did not dwell upon how the endrin got into the water, it mentioned the Velsicol plant in Memphis and also observed that in 1963 farmers applied

1.7 million pounds of endrin, 95 percent of which was used in Arkansas, Louisiana, Oklahoma, and Mississippi. In 1963 farmers in the Mississippi River basin applied endrin to 949,600 acres.[7]

Both Velsicol and Shell quickly denied that endrin had killed the fish, and they sent representatives to testify before the USDA public hearing that had been arranged by Nyle Brady. Bernard Lorant, vice president of research at Velsicol, argued that the fish symptoms were not attributable to endrin and that the PHS had neglected "fish dropsy" as a cause of death. The PHS had labeled endrin as the culprit, Lorant suggested, "only because of some analyses hurriedly made in a crash project." He preferred to link the deaths to such "invisible factors" as salt water intrusion, pollution, or a vague fish disease. Shell representative Sumner H. McAllister demanded that the benefits of pesticides be weighed against their drawbacks and pointed out that there had been fish kills along the Mississippi River since recorded time. Parke Brinkley, head of the National Agricultural Chemicals Association, recommended additional research and insisted that no action be taken until all the facts were analyzed. He also pointed out that two-thirds of the world's population was undernourished and that world hunger would only increase unless pesticides were used in agriculture. The crisis provoked editorial comment from both the *New York Times* and the *Washington Post,* which stressed the lack of federal agency coordination and posed grave questions of pesticide risk.[8]

In mid-April another round of USDA hearings in Memphis provided a propesticide forum, in which two environmentalists faced twenty-eight pesticide supporters. Witnesses from the Cotton Council and the Delta Council predicted that banning chemicals would damage the farm economy, Velsicol witnesses denied that endrin caused the fish kills, and an ARS spokesman pointed out the difficulty of banning pesticides. On the other side, Dr. James R. Whitley, who worked with a conservation group in Missouri, testified that endrin runoff from tobacco spraying had killed fish in nearby lakes, and John M. Stubbs, representing the Tennessee Game and Fish Commission, called for a ban on endrin, aldrin, and dieldrin. For the ARS, the hearing offered the opportunity for its primary supporters to speak in support of chemicals and for a few environmentalists to register their opposition. Nyle Brady could report to Secretary of Agriculture Orville Freeman that the hearings revealed no compelling reason to ban endrin.[9]

Senator Abraham Ribicoff began his April 15, 1964, hearing with a tribute to Rachel Carson, who had died the day before. Her death, he began, "brings sadness to all mankind, for she dedicated her life to the well-being of everything that lives." She was a scientist, government worker, and compelling writer. "The result was that this gentle lady, more than any other person of her time aroused people everywhere to be concerned with one of the most significant problems of the mid-20th-century life—man's contamination of his environment," Ribicoff concluded. After giving his own tribute to Carson, Secretary of Agriculture Orville Freeman questioned whether the endrin in the Mississippi River came from farm runoff or from the hundred manufacturing plants along the river. He pointed out that neither the USDA nor the FDA nor the PHS had the authority to investigate chemical plants. Ribicoff inquired about coordination among these agencies on pesticides. Freeman admitted that the USDA "did not know that the Public Health Service was even conducting a specific pinpointed research project on this until it was completed." He was miffed that USDA experts were shunned. "There are more Ph.D.s per square foot in the Department of Agriculture than in any other Federal agency," he boasted. Freeman admitted that there was a lack of coordination and turf jealousy among the USDA, the PHS, the Department of the Interior, and the FDA. He hedged on whether or not endrin was responsible for the fish kills and strongly suggested that the poison had come not from fields but from factories. Velsicol officials in Memphis, of course, denied that their company dumped endrin into the Mississippi River. Freeman seemed to accept fish kills as collateral damage, theorizing that in the future a value judgment might be required whether to take a fish kill more seriously than pesticide benefits.[10]

Ribicoff favorably mentioned Secretary of the Interior Stewart Udall's recommendation that persistent pesticides should be banned, asking Freeman to comment on Udall's opinion. The USDA, Freeman answered, was investigating that very question. Ribicoff persisted, "The point we make is that if we can't control persistent pesticides, and if this is the case, would you then agree with Secretary Udall and use them if they keep drifting these hundreds and thousands of miles from where they have been applied?" Freeman insisted that "we need to have a good deal more information in connection with both the drift, the direction, and the persistence." He then returned to weighing the positive and negative aspects of pesticides. Without pesticides, he reiterated, "we would not be enjoying many, many things in terms of our own standard of living." Ribicoff suggested that the fish kills were a warning about pesticide toxicity, and he wanted to explore the larger implications. "But if it is a question of matching off a limited fish kill because of exceptional circumstances in terms of the national nutritional level and general economic well-being," Freeman hedged, "we may have a value judgment to make." Ribicoff countered that top leaders from the involved agencies should investigate all fish kill ramifications. "I don't think we as a nation can afford, especially after the warning of the Mississippi fish kill," the senator insisted, "to sit by and wait for human beings to die." The reason that agencies moved so fast to target endrin as the killer, he argued, was because "it was a symptom of a potential danger to man."[11]

After appearing before Ribicoff's committee on April 15, Freeman returned to his office and wrote a memorandum for departmental files. It disturbed the secretary that Ribicoff had steered questions away from endrin to the larger question as to why the USDA approved toxic chemicals at all. In a rare moment of bureaucratic introspection, Freeman mused that if certain chemicals were "toxic and potentially dangerous to human beings why is any of it used at all which places the burden of proof on us." Freeman worried about the thoroughness of the ARS's evaluation of scientific studies in approving labels. "I am concerned," he admitted, "that we perhaps should have barred dieldrin, endrin, etc., because it is clear that they are toxic and, therefore, why should we use them." But he was also worried that farmers had no adequate substitutes. If the USDA banned these chemicals, as had the United Kingdom, he reasoned, "it would then place the responsibility and the burden for a crash program on the chemical manufacturers." Then he added, "Shouldn't it be there?" Basing a decision to ban chemicals on their effects on fish and wildlife was one consideration, he admitted, but harm to humans was another. Unlike his ARS administrators, Freeman anguished over persistent pesticides and the USDA's role in protecting the public, but he clung to his value-judgments argument.[12] Significantly, Freeman did not mention that if endrin used on farms killed the fish, the USDA's registration process might be compromised or doomed. In that sense, the waste produced by the Memphis Velsicol plant was a gift to the USDA.

Freeman's vacillation and insistence on value judgments prompted a letter from National Wildlife Federation executive director Thomas L. Kimball. "It is one thing for an individual to be able to do as he pleases with his own property," Kimball began, "it is another when his actions on his property affect others in a detrimental manner." Pesticide residues in milk, wildlife, and dead fish were examples of chemicals straying beyond immediate use. "We do not accept this 'either-or' attitude," Kimball wrote. "We believe that research can develop less persistent yet more selective chemicals." He mentioned several projects that seemed promising. "We are not yet ready for anyone to make a judgment that fish and wildlife must be sacrificed on the alter of agricultural overproduction," he insisted.[13]

In mid-April, Freeman met with six staffers. Because of new detection technology, he directed the ARS's Byron Shaw and M. R. Clarkson to defend newly uncovered pesticide residues in food. Inaction, he cautioned, put the USDA in the "untenable" position of "acting illegally" and "as being in the impossible position of endangering human life." Two months later, Freeman again expressed concern about chlorinated hydrocarbon residues, especially heptachlor residues in milk. "I am concerned that we did not revoke registration of it until April 28, 1964," he brooded. The USDA knew that heptachlor applied to alfalfa and cattle feed crops in the fall of 1963 would produce residues in milk in the spring of 1964. Because the ARS hesitated to make a decision, Freeman added, "we are vulnerable I believe at least to criticism for not having acted earlier."[14] Even as Freeman saw the USDA heading into dangerous waters by its inaction, the ARS refused to change its course.

Bureaucratic drift and the United States' hesitation to ban pesticides elicited a response from biologist and conservationist Julian Huxley, who on April 15 wrote a letter to the *New York Times* praising Rachel Carson. He had written the introduction to the English edition of *Silent Spring.* "I can also assure Americans that the situation in Britain at that time was as grave as in the United States," he declared, lamenting the loss of "our songbirds and birds of prey, our butterflies, and bees, and our prized wild flowers." When he mentioned this to his brother, Aldous Huxley, the novelist had replied, "We are destroying half the basis of English poetry." Julian Huxley warned that the chemical fight against insects was a losing battle, for "a remnant will always escape; still worse, new pesticide-resistant types will always develop and spread." He challenged chemical company executives who argued that judgment and action should await further study, contending, "I should have thought that the more sensible course would have been to delay the mass use of known toxic substances liable to drain into the river until more information was available as to the probable consequences on fish and other aquatic life."[15] Huxley's eloquent and commonsense evaluation of the crisis contrasted with the blunt and strained denials and self-justifications put forth by chemical companies and the ARS. In retrospect, Huxley's prescient logic appears wise, yet at the time it conflicted with agribusiness insistence that no evidence was compelling enough to ban pesticide manufacturing.

Intent on clearing farmers' use of pesticides as the fish kill culprit, on April 6 the ARS sent four teams to investigate "all plants manufacturing, formulating, or blending endrin, aldrin, or dieldrin in the Mississippi Delta south of Memphis." While some plants were either too small or too isolated to pollute the Mississippi River, several were suspect. The Memphis Velsicol plant, one team reported, released its waste into an industrial sewer that connected to the city's sanitary sewer and went untreated into the Wolf River and from there to the Mississippi River. PHS sludge samples from the industrial sewer contained up to 70,000 ppm (parts per million) endrin. Before construction of the industrial sewer two years earlier, Velsicol waste went into the sanitary sewer or directly into Cypress Creek, which ran into the Wolf River. The city dredged Cypress Creek and lined it with concrete. To remove the sludge's foul odor, city workers shoveled it atop the newly laid concrete creek bed, hoping to flush it downstream. When the flow did not remove the offensive sludge, the city hauled it to the Hollywood Road dump, which, the report observed, "is located on a loop of the Wolf River and is subject to frequent inundation." Velsicol had been dumping about a dozen drums of solid or semisolid waste there each day for six years. The PHS team took photographs and measured the chemical content of the barrels and sludge. The Hollywood dump closed in 1967; after tests revealed chemical contamination, a federal cleanup began in 1981. A July 1964 PHS investigation discovered that Velsicol discharged thirty-three pounds of endrin and nine pounds of dieldrin each day into the Wolf River. Its report estimated that "8,359 pounds of endrin and 1,089 pounds of dieldrin might be deposited in the 17,698 cubic feet of sludge" near the Wolf River.[16]

Samples taken at the Hollywood dump revealed drums with 62, 73, and 76 percent organic chloride; red liquid leaking from one drum registered 68 percent of the toxin. Sludge from Cypress Creek showed over 3 percent organic chloride, sludge from the city trucks showed 2.6 to 6.1 percent, and samples of solids from the city dump read from 70 to 74 percent. In early January 1965, Memphis spokesman George Putnicki announced that the city had discovered 8,000 pounds of endrin caked in deposits up to three feet deep in a 3,400 foot stretch of the city's sewer. The sewer was sealed

Containers of industrial plant waste disposed at the Hollywood Road city dump, Memphis, Tennessee. Part of the dump is inundated at times of high water on the Wolf River or under conditions of heavy rain.

Insecticides, box 4132m, GCV 1906–75, SOA, RG16, NARA.

off. After denying that the endrin posed any danger to Memphis residents, Velsicol in July 1965 agreed to pay $25,000 to clean up the sewer. Yet Velsicol continued endrin production, and a steady trickle of organic chlorides seeped into flood-prone dumps and streams that emptied into the Mississippi River.[17]

In mid-May 1964, Byron Shaw reported to Secretary Freeman the ARS position on the Mississippi River fish kills. While scientific findings and testimony had isolated endrin as the culprit in the fish kills, Shaw stressed "the serious consequences for both the consumer and producer if endrin, aldrin, and dieldrin should be taken off the market." The fish kills came not from pesticide runoff from agricultural lands, Shaw claimed, but from manufacturing plant waste. "In brief," he decided, "evidence obtained from both the hearing and from our own field studies does not support cancellation or modification of registration of endrin, aldrin, or dieldrin."[18] While the ARS shamelessly defended the use of dangerous pesticides—which, it claimed, had only benign results on fish and wildlife—Shell and especially Velsicol defended their plants' waste disposal policies as environmentally pure. Common sense suggests both farmers and chemical companies were the sources of pollution.

As the press publicized Velsicol's crass disregard for fish and wildlife, reporters remembered other unsavory incidents at the Memphis plant, which had been built in 1951. In 1954 an explosion in the pesticide division injured four men; a year later another explosion "wrecked the insecticide unit of the plant, killing two engineers. Fourteen others were overcome by fumes." In June 1963 Memphis health officials had received complaints of nausea, vomiting, and watering eyes from twenty persons who lived between Cypress Creek and the Velsicol plant. Residents had witnessed vapor rising from the stream. "Endrin could not have caused the symptoms," a Velsicol spokesman bluntly announced. Four days later, twenty-six workers from plants near the Velsicol factory were rushed to hospitals, sickened by chlorine gas fumes when a gas cooling tank malfunctioned. Forty persons claimed injury and twenty-six filed lawsuits. Realizing that it had both public relations and legal problems, Velsicol invited 150 Memphis politicians and business leaders to a dinner. "It came as quite a shock to us to discover that there was some question about whether we were welcome in the City of Memphis," whined John Kirk, executive vice president of the firm. Mayor Henry Loeb assured him that the plant was "very much wanted by Memphis." The city responded to its corporate friend by allowing Velsicol to dispose of its waste in public sewers, rivers, and the city dump. Velsicol continued to produce endrin, but across the river in West Memphis a PHS team monitored the river water for residues.[19]

As the press focused on the endrin story, the government warned civic leaders in Memphis—as well as in Arkansas, Louisiana, Mississippi, and Tennessee—that large amounts of endrin entered the river at Memphis. Velsicol boldly defended its endrin production and denied culpability. Louis A. McLean, the secretary of Velsicol, questioned how endrin entering the river in Memphis was killing fish in Baton Rouge but not in between. "Entries of pesticides do not first create fish deaths 500 miles away," he asserted. Velsicol's Bernard Lorant flatly contradicted PHS research. "Endrin is not the causative factor in the Louisiana–Mississippi River fish kills," he stated. Arguing that endrin runoff was impossible, Sumner H. McAllister, the general manager of Shell's chemical operations, claimed that experiments showed that mud acted as "a natural purifying agent" that would remove endrin before it reached the river. On June 16, Senator Everett McKinley Dirksen of Illinois defended Velsicol, a Chicago-based corporation, calling the PHS studies "wild accusations."[20]

Senator Abraham Ribicoff concluded that the endrin readings at Velsicol's Memphis plant indicated that at least one source of endrin contamination had been found, but he suggested that there were others as well. A witness before his committee, he recalled, had seen dead fish in a bayou an hour after a field had been sprayed with endrin. At a conference in New Orleans in May 1964, K. E. Biglane, a Louisiana biologist, witnessed endrin application and returned after a rain to find "many thousands of fish and snakes, turtles, eels dead and dying." Fish were dying in Bayou Teche, where the town of Franklin got its drinking water. In early July, six Louisiana bayous had substantial fish kills following heavy rains, and endrin was suspected as the toxic agent.[21] Dismissing the stonewalling of both chemical company executives and ARS apologists, Ribicoff concluded that the reckless manufacture and application of endrin was causing a major environmental crisis.

The fish kill issue continued to garner headlines and to provoke congressional interest throughout the summer of 1964. The PHS report charged that both the Memphis Velsicol plant and agricultural runoff contributed to the fish kills. The USDA countered that there was no evidence that agricultural runoff was to blame. ARS head Robert J. Anderson admitted that there could have been "local kills" due to improper use of endrin, but not "chronic kills." Anderson had studied veterinary medicine at Texas A&M and had worked for years in the Bureau of Animal Industry before moving to the ARS in 1961 as assistant deputy administrator for regulatory and control programs. Meanwhile, Velsicol continued to deny that its Memphis plant was responsible for the fish kills.[22]

To avoid any more problems with the city, in 1965 Velsicol opened a 243-acre facility in Hardeman County, Tennessee. It proceeded to dump a quarter million 55-gallon chemical residue drums there between 1965 and 1972, when complaints from local residents shut it down. The county health department determined that local wells had been polluted and feared that both the Hatchie River and the water table furnishing water to Jackson and Memphis contained pesticide residues. Woodrow and Christine Sterling, who lived half a mile from the dump, chronicled a number of family health problems. They appealed without effect to local doctors and to the Environmental Protection Agency regional office in Atlanta. Their well contained traces of carbon tetrachloride, chloroform, benzene, heptachlor, chlordane, and chlorobenzene, all suspected carcinogens. In 1978 Velsicol agreed to pay for water supply connections for twenty-six homes near the dump. "A possible seepage was found in what was previously considered by experts a secure landfill," a Velsicol spokesman admitted. Ultimately Velsicol paid $12 million to residents who lived near its Hardeman County dump. After consulting with the Center for Disease Control and a Vanderbilt University toxicologist (presumably Wayland J. Hayes Jr.), Velsicol's medical services division decided against doing liver biopsies. Velsicol's plant manager stated that after the Hardeman County dump closed, the company was "recovering or reusing most of the waste now and that the remainder was being burned in an incinerator at the company." Fines, judgements, and lawsuits did little to blunt Velsicol's propensity to violate the environment. Its dubious legacy continues to this day. A 2003 study focused on the Cypress Creek area near the Velsicol plant to determine if toxic residues persisted. Soil tests revealed dieldrin, chlordane, endrin, and heptachlor, and the Tennessee Department of Environment and Conservation put up fences near Cypress Middle School to keep people away from the contaminated area. Studies were underway to determine if people who lived along Cypress Creek had elevated cancer rates. Residents claimed that chemical residues destroyed property values as well as their health, and in 2004 a $1.75 billion lawsuit against Velsicol was pending.[23]

Three years after the highly publicized fish kills, ARS investigators continued to examine waste from chemical firms. On April 3, 1967, W. F. Barthel, the ARS chemist for the southern region, met with a Velsicol official and warned him that a recent survey had revealed chemical waste "hot spots." ARS head Robert J. Anderson, Barthel informed the official, "felt that his company should have a chance to look over the data so that it might take any corrective action prior to the report being released to other agencies." The official met Barthel and other ARS staffers at Gulfport on April 6 to examine both 1964 and 1966 reports on pesticide readings. He did not challenge the data. Barthel also thoughtfully advised a Velsicol official that sampling would soon take place in the Wolf River and in Cypress Creek near Velsicol's Memphis plant.[24]

The ARS combined monitoring Mississippi Delta chemical companies and pampering corporate executives. On April 12, 1967, Barthel met in Canton, Mississippi, with Dr. John Connors of the Champion Chemical

Company and a Dr. Blaylock of the Riverside Chemical Company. The men expressed no surprise that Barthel found residues near their plants, and Connors "freely admitted that they were dumping their waste at the stream site where the residues were found." At Barthel's urging, Connors called a meeting for April 28 in Greenville, Mississippi, which Barthel assured him would be "a closed meeting, only formulators are to be present; no newsmen or publicity." If ARS inspectors approached factory managers or corporate executives without a preliminary meeting, Barthel suspected, they might feel "singled out" and "refuse us permission to enter their plants" or "relegate us to some minor official who had no power to make decisions."[25] Given this deference, Barthel appeared entirely willing to allow the chemical companies to dictate ARS investigations. Surely his warnings of impending ARS visits allowed chemical companies the opportunity to clean up as much as possible in advance. It was unclear what, if any, action the ARS would have taken had it acted on any violations its inspectors discovered.

Despite the warnings, ARS investigations revealed disturbing conditions. On May 4, 1967, an official escorted inspectors around the Niagara Chemical Plant in Greenville. Waste from the plant flowed into a pond some five hundred feet from the factory. The plant used "a large old boiler" to burn bags, boxes, and other trash. The ashes were buried "on their own fenced property in slit trenches." Eventually, the plant would run out of land and would need to dig additional trenches where the original waste had been buried. Barthel wondered if seepage would contaminate nearby well water or if ditches would carry polluted water into nearby Fish Lake. "It is interesting," he reported, "that the six samples taken from Fish Lake area in 1966 all showed DDT family residues. Only one of six in 1964 showed residues." The ARS sent a copy of the report to the manager of the Niagara Chemical Company in New York, but there is no indication that any action was taken.[26]

Velsicol and Niagara executives requested copies of the ARS's 1967 reports. "I have instituted many changes in our operations and have noted substantial improvement in the past few years," S. H. Bear of Niagara explained to Robert Anderson. "There is room for much more. Our efforts will continue." Bear insisted that industry could solve its problems, writing, "I oppose the view that government legislation and regulation are mandatory." He admitted that unless the corporate world became more sensitive to the environment, federal action would follow. Then Bear wrote several cryptic lines. "U.S.D.A. can help. We need a continuing exchange of views. We may need some 'hidden persuaders.' If, through legislation, you get the job, you will need our help."[27] The words lend themselves to various interpretations, but they unmistakably show that the corporate world knew its friends. Bureaucrats could protect corporations from government rules, and down the road payback time would come.

Chemical companies disregarded the impact of lax waste disposal and the health of workers. Whatever results long-term epidemiological studies on chemical company workers might reveal, conventional wisdom contains cautionary stories. Attorney John McWilliams grew up in Holly Ridge, Mississippi, and his family owned a general store where laborers would congregate for lunch. In the mid-1960s, he recalled, Mission Brand Chemicals opened a plant nearby and employed young black men in their late teens. These workers arrived for lunch at the store covered with white or yellow dust, "and you couldn't tell that they were black." McWilliams allowed that from their appearance, "they must have jumped down in the bin and mixed this stuff up by hand." He mentioned several by name. "And they all are dead," he continued. "They died young." McWilliams's law partner, Lawson Holladay, recalled a summer job he had while in college spraying the vegetation along the Sunflower River with "2,4-D, 2,4,5-T, diesel fuel and a soap surfactant to make it stick. I mean, that's Agent Orange." After discussing the impact of pesticide spraying in the Mississippi Delta, Holladay observed that "after they finally banned DDT, we do have some bald eagles nesting in the next county."[28]

Major fish kills declined, but endrin and other chemicals continued to kill fish. On July 26, 1973, an aerial applicator in Alabama sprayed endrin on fields near A. R. Burroughs's pond, which he had built in 1965 and stocked with game fish. Fish started dying the day after the spraying, and, as the court decision summarized, "the fish continued to die until all of the fish in the pond were dead." When Burroughs restocked the pond, the fish did not survive. A South Carolina cotton farmer had his crop sprayed with 6-3 and cotoran on July 10, 1973. George R. Green, owner of an adjoining eight-and-a-half acre fish pond, discovered dead fish, and when tests revealed pesticide residues in the water, he sued the farmer and aerial applicator.[29]

In 1979 a study targeted endrin as a cause of brown pelican die-offs in Louisiana. In the fifteen years before 1933, pelican counts varied from 12,000 to 85,000, but the population crashed in the late 1950s and the last

A healthy pelican egg beside an egg weakened by chlorinated hydrocarbons.

From Science in American Life exhibit, National Museum of American History.

survivors in Louisiana nested in 1961. The specific cause of the extermination of brown pelicans in Louisiana was not clear, although their position at the top of the food chain in a highly toxic area seemed a likely cause. The die-off also coincided with the fish kills and the endrin releases from the Memphis Velsicol plant. Pelicans from Florida were introduced into Louisiana, but, according to a 1979 study, die-offs in 1975 "coincided with the peak in endrin residues in pelican eggs." The pelican diet of fish loaded with chlorinated hydrocarbons led to weakened eggshells and reproductive failure. After endrin was banned in 1984, brown pelicans again flourished in Louisiana.[30]

As chemical companies and ARS staffers weighed their options, the effects of the chemical revolution were working themselves out in farmers' fields and in streams and rivers. "This is the season in the cotton country," reporter Roy Reed began a December 1969 article, "when the air turns clear and sweet and when the fish sicken and die." Many of the fish that did not die, Reed continued, had built up a resistance to chemicals and stored vast amounts in their bodies. They had become "living bombs" to any animal eating them. The problem came with technological changes in agriculture, Reed explained. Mules yielded to tractors and hoes to herbicides, and workers left the land. When the minimum wage was applied to agricultural labor in 1967, the price for labor rose from $3.50 a day to $1.30 an hour, or roughly $10.50 for an eight-hour day. In 1960 labor cost Mississippi cotton farmers $46.80 per acre, while chemicals cost $13.50. Hands chopped out the weeds in the spring and summer, and some farmers were still using hand pickers in the autumn. When the minimum wage kicked in, farmers turned to chemicals. The cost of labor accordingly dropped to $13.50 and chemical costs rose to $24.00. Reed explained that insects built up resistance to chemicals, compelling one farmer he interviewed to spray nine times per year instead of three. Insecticides intended for boll weevils also killed insects that preyed on the bollworm, creating other major pests. In the fall, when defoliants were sprayed, the smell and residues drove people inside. Dr. Denzel B. Ferguson at Mississippi State University found small mosquitofish in the delta loaded with 120 times as much toxic chemicals as once would have killed them. When he fed a healthy but pesticide-loaded mosquitofish to a cottonmouth, the snake died. In a paper that Ferguson gave in 1967, he made the disturbing observation that "the time may come when we will be more concerned over the fish that survive a pesticide kill than over those that are killed."[31]

6
POISONING BY DESIGN

All the screaming and hollering hasn't affected sales, so it's mostly a PR war at this point and there isn't enough info to base any substantive action.

George Mehren

Silent Spring, rising environmental consciousness, residue studies, poisonings, lawsuits, and fish kills generated tension in what had been a chummy relationship between the USDA and the chemical industry. Mindful of its responsibility in monitoring pesticides, the USDA straddled environmental pressures and corporate demands, proffering rhetoric to please both sides. Parke C. Brinkley, president of the National Agricultural Chemicals Association (NACA), prided himself on the organization's close ties to the USDA, but the department's measured statements on *Silent Spring* and on pesticide safety upset him. He was further provoked by a syndicated editorial column by Rowland Evans and Robert Novak, who claimed that Secretary of Agriculture Orville Freeman had forced ARS deputy administrator W. L. Popham into retirement because of his propesticide posture. Along with a blistering letter to USDA undersecretary Charles Murphy, Brinkley enclosed an article from the June 1964 issue of *Agricultural Chemicals* that branded the Evans and Novak editorial "an unfair and unwarranted slur on the record of a dedicated career scientist," pointing out that Popham had announced retirement plans "some time ago." The article insisted that the USDA's failure to comment on the Evans and Novak story suggested that it was accurate. Popham's treatment, Brinkley wrote, "gives me a great deal of concern," and he scribbled across the clipping, "Will Dr. Clarkson get the same treatment?"[1] Popham and ARS deputy administrator M. R. Clarkson had enthusiastically supported the chemical industry, and Brinkley understood who in the ARS best served the interests of agricultural chemicals.

Brinkley's letter and the enclosed clipping prompted a spirited discussion within the USDA. The department had not commented, public relations staffer George Barnes explained, because "Dr. Popham's interest

Parke Brinkley.
Compliments of CropLife America.

would be best served by ignoring the story." But Brinkley's attitude epitomized a larger problem between the chemical industry and the USDA, Barnes confided to another staffer. "Mr. Brinkley is mad at me (and the Department) currently," he continued, "because he feels that we have not defended the chemical industry with sufficient vigor in the pesticides panic." It was hardly the time, he had argued to Brinkley, for Secretary Freeman and agricultural chemical executives "to indulge in public displays of togetherness." Under the circumstances, the USDA was doing all it could to protect the chemical industry. "What Mr. Brinkley overlooks," Barnes offered, "is that the Department, while not rushing to the ramparts, has steadfastly held the line on the use of pesticides against very heavy pressure of public opinion." The USDA had cancelled the registration of only one chemical, heptachlor. "In this instance, we had no alternative," Barnes admitted. In the case of endrin, the culprit in the fish kills, "we have announced that we are not going to withdraw registration." Brinkley had also overlooked Freeman's prochemical remarks before congressional committees. "I made it clear, I thought, that we were in no sense shunning the industry," Barnes concluded.[2]

At some point, Barnes suggested, the secretary of agriculture "may find it appropriate and necessary to make an all-out public review and defense of pesticides—but in my judgment, the proper forum for such an exercise would be, say, the League of Women Voters or the National Farmers' Union—not the assembled chemical manufacturers, with whom a not inconsiderable body of public opinion already links him in an unsavory way." The problem was, in essence, one of public relations. "While we do not want to embrace the industry we do not want to antagonize it, either," he recommended to Freeman on May 26, 1964. Barnes had learned that Brinkley was furious that the USDA had not attacked Rachel Carson's charges "that county agents and others were the servants of the chemical industry"; moreover, Brinkley had implied that the USDA "had been weak-kneed." Revealing chemical industry tactics, Brinkley had admitted to Barnes that the NACA could not act "directly" but instead was "working through 'other people'—scientists, state college specialists, and so on." Barnes had turned down an offer to make a NACA-sponsored tape for five hundred radio stations and had found Freeman's schedule too tight to fit in a meeting with the head of Shell Oil's agricultural chemical division.[3] While Brinkley obviously thought that the USDA should be a public relations arm of chemical manufacturers, the USDA attempted both to shield itself from public contact with the pesticide industry and to promote pesticides. Even as the USDA publicly hedged on chemical use, Brinkley's "other people" continued to attack Rachel Carson and to lobby for agricultural chemicals.

By publicly distancing itself from pesticides advocates, the USDA incrementally moved away from its stereotype as the handmaiden of the chemical industry. Citing an article on chemicals in late 1965, USDA staffer Harold Lewis complimented Secretary Freeman for his public relations posture. "We have had relatively little trouble in the last year," Lewis boasted, "and I think if we did what the pesticide industry would like to have us do in supporting its case publicly, we would tear down a good deal of positive reaction which has begun to develop relative to agriculture and its use of chemicals." Deputy ARS administrator George Mehren wrote on the bottom of the memo, "Yes. This has been my position all along. All the screaming and hollering hasn't affected sales, so it's mostly a PR war at this point and there isn't enough info to base any substantive action."[4] The public relations posture no more addressed faulty labels, falsified data, ineffective products, or safety concerns than it did the growing worries that agricultural chemicals created more dangers than benefits.

Despite some friction between industry and the ARS, in most cases they worked together closely—some might judge too closely. In the summer of 1967 Parke Brinkley suggested that the ARS host a regulatory workshop in early September. On July 18 he had met with Harry W. Hays Jr., director of the Pesticide Regulation Division (PRD), and had worked out a format. Industry representatives would present their problems, and USDA, FDA, U.S. Fish and Wildlife Services (FWS), and Public Health Service (PHS) staffers would present their views and recommendations. There would be ample time for general discussion. Having the people who wrote and enforced the rules meet with the people who applied for registration, the reasoning went, "would be of great benefit to both the government agencies and industry." Brinkley insisted that Harry Hays head the panels. He envisioned about a hundred people attending the workshop.[5]

In addition to sponsoring genial workshops, the agricultural chemical industry continually pressured the USDA to support its agenda. In February 1969 the agribusiness and public affairs firm Donald Lerch & Co. set up a meeting of industry representatives with USDA officials to discuss "the urgent need for the Department to give higher priority to handling the pesticide residue problem in Europe and to find a new way to contain it." Both Lerch and Brinkley feared that the United States would lose part of its European market if an approaching European community meeting lowered residue tolerances below those in the States. The two government representatives, one from the ARS and the other from the FDA, "are extremely busy," Lerch argued, "and our point is both more attention and new approaches must be found or the U.S. will pay the price in lost sales due to excessive residues." Lerch and Brinkley would be accompanied to the meeting by "six men personally familiar with the situation," who were "in a position to offer constructive suggestions." The men represented Hercules, Shell, Velsicol, Union Carbide, Stauffer, and the NACA. Lerch thoughtfully suggested three suitable representatives from the USDA.[6] The pesticide lobby judged the issue too important to be left solely to USDA bureaucrats. By using muscle, chemical interests could shape USDA policy and, instead of reducing pesticides in U.S. products, insist that Europeans accept higher pesticide residues.

Increased attention to pesticides called into question policies that the USDA had long accepted. When Secretary Freeman refused to speak before the NACA convention, he implicitly indicated that he was fighting an image problem. Others in the department also reacted to scrutiny. "Throughout the pesticide controversy," Nyle C. Brady confided to three of his colleagues, "writers not friendly to Agriculture have insinuated that we are under the thumb of the chemical industry and that our programs reflect the primary interest of the industry." Critics focused on the close association between industry and land-grant colleges, especially in industry-funded projects. Brady was curious how much financial support industry gave to land-grant schools. "Some stations received no support in some years and others a very low level of support," T. C. Byerly vaguely replied. In contrast, some schools received over $200,000 a year. He estimated the total at between $1.5 and $2 million annually. Schools also received in-kind support, primarily in chemicals and supplies.[7] This was obviously a sensitive question, and Byerly danced around it. He suspected that a close examination of the relationships between industry and land-grant schools would suggest collusion.

The ARS also maintained close ties with Congress, especially with Mississippi's powerful congressman Jamie Whitten. In October 1969, Ned Bayley briefed Whitten on the implications of restricting organochlorines on cotton, corn, peanuts, and tobacco. In 1966, 87 percent of organochlorines, some 72 million pounds, were used on those four crops. By substituting 42 million pounds of organophosphorus and carbamate pesticides for organochlorines, Bayley estimated that farmers could "still maintain effective insect control on the four crops." Corn and cotton would still require organochlorine applications. The changeover would cost cotton farmers $15.4 million and peanut farmers $1.4 million, and increased costs for all four crops would average $2.23 per acre. The change, Bayley extrapolated, would mean some restructuring in the chemical industry, a reduction in environmental pollution, additional threats to bees, a small rise in consumer prices, and increased health risks to the people who applied organophosphates. To Parke Brinkley, restricting chlorinated hydrocarbons was another slap at the chemical industry. "The pesticide industry is presently operating in a very unfriendly, if not hostile, public climate," he grumbled to agriculture secretary Clifford M. Hardin in May 1970. Companies were cutting back on research or simply closing their doors.[8]

Even as the ARS coddled the chemical industry and did little to police organophosphates, it attempted to offset criticisms by suggesting a future with benign control methods. "Our total experience up to this point indicates that a realistic goal for the immediate future would be to aim for 90% insect control by other means and 10% by chemicals," ARS administrator George W. Irving Jr. claimed in a May 1969 New Orleans speech. NACA president Parke Brinkley bristled at this suggestion. "I think we are going to have to depend very heavily on chemical pesticides for the foreseeable future," Brinkley remonstrated to Irving in June 1969, and he strongly recommended added research funds be directed to chemical controls. "We are, however, having a tremendously difficult time in combating the people who would take away from the farmers of this country the production tools that they so badly need in keeping the cost of production down," Brinkley grumbled, "and statements such as yours certainly hurt the cause." By shifting the question to economics and farm needs, Brinkley dismissed concerns about chemical residues and poisonings and portrayed chemical companies as besieged patriots. He linked Irving to "the people" who opposed pesticides.[9] Irving certainly knew what that meant.

Unsure of his statistics and even puzzled why he had made the statement, Irving immediately began damage control. He ordered a staffer to check his recent speeches and testimony for "a statement to the effect that we will be using chemical pesticides for a long time." He wanted to respond to Brinkley, but, he insisted, "I don't want this reply to be defensive." Ultimately, Irving stressed "the current furor over DDT" and stoutly added that "the public will not accept pesticides that it considers dangerous to the environment." He praised E. F. Knipling's work with sterilized insects but could not predict when such nonchemical controls would become practicable. Farmers would continue to use chemical controls until something better emerged, he concluded, but he defended his remarks on nonchemical methods of control. "I think instead that we are building confidence in the agricultural community," he hedged, "by showing that we won't always have to depend *solely* on chemical pesticides to ensure our food supply."[10] He implied that nonchemical research was more linked to public relations than to insect control.

Irving's earlier statement had unintended international ramifications. M. B. Green, who headed a pesticide group in London, requested permission to print part of Irving's New Orleans speech. This set off a series of memos within the ARS, which resulted in Irving requesting that Green change his abstract to reflect the revisions he had made to please Brinkley. It should read, Irving advised, that the USDA goal was "to aim for 90 percent insect control by alternate means including highly specific chemicals and 10 per-

George W. Irving Jr., 1947.
Photo by Forsythe. 16-N-8098, NARA.

cent by conventional chemicals."[11] This sufficiently diluted the statement to justify the status quo.

Irving's discomfort over Brinkley's pressure derived in part from *Silent Spring* and Rachel Carson's subsequent congressional testimony, which resonated with the experiences of people who detested pesticide drift. Californian C. J. Robben located a crucial set of contradictions in USDA policy. Farmers intent on making a living, Robben wrote in August 1965, "are perhaps substantially poisoning their families and habitat." He charged that farmers were "dupes of certain powers within the Department of Agriculture," who every year "accept less insect damage and more 'pesticide' residue." Robben suggested that the increasing use of pesticides originated "either direct or indirect, from the chemical works which produce the pesticides." As he looked around his Dixon, California, community, Robben pondered why he and his neighbors were compelled to share pesticide exposure with insects. "Being directly sprayed and bombarded as often as once or twice a week and sometimes every day for a week during a period of months—like something from World War II 'gas warfare' is no joke," he declared, asking if there were rules to control such spraying. "Also, let's face it," Robben wrote in a devastating observation, "a little insect damage to the agriculture products we consume, while perhaps not looking as pretty, will do us substantially less harm than the consumption of poisons used to kill them." In its nonresponsive reply to Robben's serious questions, the ARS sent him "picture story No. 168," which described the labeling of pesticides.[12]

To Jane and James Ward, chemicals seemed inescapable. They had left "big-city air pollution" for the "wide-open spaces of Texas," the couple informed Senator Ralph Yarborough in June 1969. Instead of pure air, they fretted, "we find ourselves with the choice of breathing malathion from the mosquito-fogger in town or, in the country, smelling on every sweet breeze the drift of parathion from the sorghum fields." A new cotton herbicide would be used in the fall, they had learned, and the USDA was applying dieldrin to large parts of nearby Kelly Air Force Base in San Antonio, which the couple deemed "one of the more flagrant and insane atrocities." In Woodson County, Kansas, G. W. Gandy and his wife owned twenty-three acres and were nearing retirement age. "Some of the farmers had their pastures sprayed for weeds and other things the last week of May," they wrote on July 16, 1969. "Since then most of our fruit trees have died or are dying, the rest of the garden has been heavily damaged, even the shade trees have been injured." On the day of the spraying, they complained, "the fumes and smell were bad, we had to go inside for our own protection." They wondered if there were no laws to protect people from such exposure, pessimistically concluding, "If not then we might as well give up."[13]

Farmers relied upon advice from chemical salesmen, merchants, county agents, and farm periodicals, as well as word of mouth. Because they did not always abide by label guidelines, users exacerbated pesticide dangers. To cope with a wet planting season in 1967, Arkansas soybean farmers used arsenical herbicides and arsenic acid to clear fields of weeds before planting. "As you know," research agronomist Roy J. Smith explained to his superior, "arsenical herbicides are not registered for use in soybeans," and he worried about such formulations being used "in an illegal way." In addition, drifting herbicides had damaged nearby rice fields, raising questions about "harmful residues" in harvested rice. Even the USDA suffered cattle losses when, in the summer of 1967, workers at the Goddard NASA facility (which abutted the USDA's Beltsville farm) ignored their neighbor's advice and sprayed sodium arsenite along its fence. Nineteen USDA cattle were killed. The cattle corpses, containing 300 to 400 ppm arsenic, were sent to a renderer in Hyde, Maryland.[14]

In May 1968, Alabama's two U.S. senators complained of crop damage following herbicide spraying in Bankhead National Forest. Seventy-two farmers petitioned for damages, and over 150 people attended a public meeting about the issue. The damage extended some nine miles from the spray area. A month later ARS investigators reported that the cotton plants had "completely recovered" and that the damage came not from herbicide drift but from a "very heavy infestation of thrips during early stage of plant growth." Strangely, the thrip infestation occurred only in the area hit by drift. Such self-serving reports raised as many questions as they answered.[15]

The ARS not only monitored the press but sometimes pressured writers. When in December 1966 *Progressive Farmer* staffer Bill Barksdale sent an advance copy of his article, "Would You Poison Your Well?" to the ARS for review, agricultural engineer Robert F. Eagen suggested that Barksdale eliminate his first paragraph, revise the second, and make changes in yet another. Unless these changes were made, Eagen warned, the article "will tend to alarm readers unduly, considering the extremely low incidence of pesticide-related accidents in the past." Barksdale agreed to the revisions.[16] Such control over the rural press allowed the ARS to minimize the negative effects of pesticides and to present a wholesome image. Given the actual incidence of poisonings, readers would have profited from accurate reports of pesticide toxicity.

Enthusiastic articles and glossy advertisements did not work with everyone. In July 1969, Stanley J. Medine of White Castle, Louisiana, complained to Secretary of Agriculture Clifford Hardin of a two-page Shell Chemical advertisement that depicted rolling fields of corn with an impressive farmhouse in the background. The text of the ad stated that such fields were a "smorgasbord" for insects, which ate a third of the crop. To stunt the appetite of nematodes that were eating some half billion dollars of all food crops, Shell recommended Nemagon and Gardona. Meanwhile, Shell scientists were busy doing research on other chemicals. "Because today," the ad continued, "with the world's population zooming, America's 'amber waves' of grain are too precious to be wasted on greedy bugs and worms." Medine was disturbed especially by the last sentence. The idea that chemicals were presented as essential for production, he pointed out, "apparently does not take into consideration the governmental fact that the USDA, thru its Crop-Subsidy and Soil-Bank Program, has a 'little say-so' as to the total allowable food production on America fertile lands." He allowed that few people believed in folklore, but he recalled a saying from his youth: "We'll plant enough for us . . . 'n the birds . . . 'n the bugs." While not poetic, he admitted, it was "a sound formula in promoting a healthy, wholesome and beautiful environment." He dedicated the letter to the memory of Rachel Carson.[17]

In June 1968, Wayne M. Blickhahn, president of the Marinette County Council of Sportsmen's Clubs, complained to his Wisconsin congressman about pesticide testing, insect resistance, birth defects, reproduction failures of eagles and ospreys, chemical residues in wildlife far from where poison was applied, slower reaction times of wildlife and humans after repeated exposure to toxic chemicals, and the general degradation of the environment. ARS director of science and education Ned Bayley evaded Blickhahn's questions and, after citing statistics on fish kills, insisted that the ARS paid close attention to wildlife issues and conducted important research on nonchemical means of pest control. Bayley promised "constant vigilance" to protect fish and wildlife. By hiding behind the claim that the ARS took a serious interest in wildlife and research, Bayley's response proved misleading and nonresponsive.[18]

As tenants, sharecroppers, and small farmers were pushed off the land and the structure of agriculture was transformed into capital-intensive operations, farmers constantly adjusted the types and amounts of pesticides they used. While statistical sources vary slightly , they convey general trends in pesticide use. In 1966 farmers used some 414 million pounds of pesticides (insecticides, herbicides, fungicides, etc.); government, industry, and homeowners increased total pesticide use to 682 million pounds. By 1971, farm pesticide use had ballooned to 578 million pounds and total use had risen to 793 million pounds. Although cotton acreage had been reduced in 1966, farmers still used 64.9 million pounds of insecticides on the crop, which comprised fully 47 percent of all insecticides applied in agriculture. In 1971 increased cotton acreage lifted this amount to 73.3 million pounds. As boll weevils and other insects built up resistance to DDT and other chlorinated hydrocarbons, farmers turned to organophosphates, the sales of which increased from 40 million pounds in 1966 to 71 million pounds in 1971. Chlorinated hydrocarbons commandeered 62 percent of the pesticide market in 1966, with toxaphene, DDT, and aldrin dominating half the market. By 1971 the gradual banning of chlorinated hydrocarbons had reduced sales. Southern farmers used more insecticides than their counterparts in other sections of the country. In 1966 and 1971, farmers in the southeast used 35.4 million and 40.4 million pounds of chemicals respectively; in both instances, this figure represented 26 percent of the total market. In the Mississippi Delta states, pesticide use increased from 21.8 million pounds in 1966 to 32.3 million pounds in 1971. Herbicide use also increased dramatically as more farm workers left the land and chemicals replaced hoes and cultivators. From 1964 to 1966, herbicide use increased 37 percent, from 84 million to 112 million pounds; by 1971 it had reached 226 million pounds, 36.5 million pounds of which were used by cotton farmers. Atrazine and 2,4-D accounted for half of all herbicide sales in 1966; atrazine sales rose from 23.5 million pounds in 1966 to 57.4 million pounds in 1971. Organophosphate use increased from 33.9 million pounds in 1964 to 40 million pounds in 1966, with parathion, methyl parathion, diazinon, and malathion accounting for two-thirds of the sales. Cattle operations, which used primarily toxaphene, methoxychlor, dichlorvos, malathion, carbaryl, and DDT, consumed 6.2 million pounds of insecticides. The Agricultural Census reported that from 1960 to 1997, the outlay for agricultural chemicals increased from $908 million to $7.5 billion.[19]

As chemical sales increased, poorly labeled toxic chemicals continued to take a toll on human health, especially that of children. Sensing that pesticides were transgressive, children were drawn to them. In the spring of 1971, three children playing in a shed near Fountain, North Carolina, found a "partially filled jug of 4 pound emulsifiable concentrate parathion." Thinking it was wine, they sipped it. The five-year-old girl died, and her brothers ended up in intensive care in a Greenville hospital.[20]

In 1967 two physicians released a study of parathion deaths in Florida that spanned a six-year period from 1959 through 1964. They reported fifty-six deaths, thirty of them children. Most of the children were under three years old. Five were white and twenty-five nonwhite. Nine people also died from malathion poisoning during these years. The children died from drinking parathion from improper containers (such as soft drink bottles), contacting it on floors or window sills, inhaling it, or purposely taking it. The study focused on the deaths of two children and the sickness of another. The three children made a swing from a discarded parathion sack and became ill on their second day of swinging. The nine-year-old girl went to the hospital at 8:30 P.M., "when she had developed high fever, choking, dip-

TABLE 1
Pesticide use, 1966, 1971, 1999

	Farm use (lbs. of active ingredient in millions)	Total U.S. (lbs. of active ingredient in millions)	Insecticides, farm use (lbs. of active ingredient in millions)	Herbicides, farm use (lbs. of active ingredients in millions)	Fungicides, farm use (lbs. of active ingredients in millions)
1966	414	682	149	112.4	33
1971	578	793	170	225.7	39.5
1999	706	912	93	428	45

SOURCES: Compiled from "Quantities of Pesticides Used by Farmers in 1966," Agriculture Economic Report No. 179, Economic Research Service (USDA, 1970); "Farmers' Use of Pesticides in 1971," Economic Research Service (USDA, 1974); David Donaldson, Timothy Kiely, and Arthur Grubbe, "Pesticides Industry Sales and Usage: 1998 and 1999 Market Estimates" (EPA, 2002).

lopia, nausea, and vomiting, abdominal pain, weakness, headache, dilated pupils, and increased salivation." A physician saw her in the emergency room. Her condition deteriorated, and she "quietly expired 1½ hours after arrival at the hospital." Her five-year-old brother arrived at the emergency room at 11:00, an hour after his sister died. He had much the same symptoms as his sister, and he died five hours later. The eleven-year-old boy "developed fever, weakness, nausea, vomiting, increased sweating, muscular tremors, double vision, and inability to walk that evening." He arrived at the emergency room at 3:00 the next morning. The physicians treated him with atropine and oxygen. After six days in the hospital, he recovered and was discharged. All of the children had low cholinesterase levels. The authors of the study strongly recommended that parathion not be used around homes and that it should be available only to entomologists or "experienced agriculturists." Incomplete ARS statistics on parathion listed three fatalities, seven "non-fatalities" and thirteen "suspects" from November 1969 through July 1970. The cases came from all over the country.[21]

One of the most disastrous parathion accidents occurred in Tijuana, Mexico, in September 1967, when seventeen children (aged between twenty months and twelve years) died and two hundred others were poisoned. The San Luis Mills in Hermosillo, which packaged pesticides, also stored flour and sugar in an adjacent room. Parathion was accidently mixed with the sugar and flour, shipped to bakeries in Tijuana, and baked into bread. One slice of bread tested 11 ppm parathion, and the stomach contents of one of the victims contained 27 ppm parathion. In a similar warehouse accident in 1952, a Georgia milling company sued a couple that bought but did not fully pay a bill for $4,134.37 for contaminated feed to grow 2,400 turkeys. The turkey feed had been stored in the company warehouse along with DDT and other toxic pesticides. The couple lost 912 of their poults, and the survivors were stunted. They filed a cross-action suit for their losses, and the jury awarded them $4,560.[22]

Many agricultural workers were constantly exposed to organophosphates as a matter of course. California workers frequently developed what was called "orange pickers' flu," which was characterized by nausea and headaches. California acted to protect workers by setting a time period after spraying when workers could not enter the fields. Cesar Chavez and the Farm Workers Union pressured growers to protect poor and vulnerable workers. Suggesting more widespread poisoning, a physician in Tulare County hinted that farm workers seldom reported illnesses for fear of losing their jobs. A study of 301 workers by the Field Foundation in 1971 revealed that over the course of a year, one third were accidently sprayed and 30 percent developed symptoms of pesticide poisoning. Chemical companies justified toxic chemicals by claiming that agricultural production would sharply decline without them. Gordon Snow, a California Agriculture Department staffer but no historian, declared that if organophosphates were banned, "the country could support about as many people as it did when the Indians inhabited the continent." Monsanto's Marshall Magner boasted that the safety record of organic phosphates "is darn good." He did not deny that there were illnesses but added, "I suspect some of the reported illness is psychological."[23]

Evidence was accumulating, however, of widespread health problems traceable to chlorinated hydrocarbons. In September 1969, T. C. Byerly summed up research on DDT. "It is likely that there will be increasing allegations of nerve and brain damage from chlorinated hydrocarbon pesticides," he notified Ned Bayley. Byerly cited Bruce Welch, a highly regarded scientist who had "asserted that all of the chlorinated hydrocarbon pesticides affect the cholinesterase system and that chronic effects may be severe." Welch and other scientists had faulted Wayland J. Hayes Jr. "for ignoring possible chronic brain injury."[24]

Hayes continued to act as a consultant for the ARS. His reaction to organophosphates sometimes seemed disingenuous. On February 29, 1972, several ARS staffers met with Hayes, who had moved from the PHS to Vanderbilt University. They discussed reports of parathion poisoning in vineyards, orchards, and groves; they were also concerned about delayed poisoning symptoms among workers in the ARS pesticide scouting program. "With continued exposure to organophosphate pesticides," Kenneth C. Walker reported, "we can expect a gradual decrease in cholinesterase in the workers." Such symptoms, he continued, were expected and could be explained to workers as a normal reaction. "They should also be told that a large drop in cholinesterase, if it should happen to occur, can be tolerated by the human body as they have a very large excess of cholinesterase," he advised. If workers' cholinesterase level fell, he suggested, they should stay out of the fields till it built back up. The physicians and ARS staffers ignored the danger of organophosphate buildup and the consequent danger of a cholinesterase crisis upon reentering toxic fields.[25] The report implied that headaches, nausea, and vision problems should be considered a normal part of working around organophosphate pesticides.

Bill Robinet, who flew dusters in Arizona, Mexico, and Canada in the mid-1950s, had vivid memories of parathion poisoning. In his autobiography, *By the Skin of My Teeth,* he narrated his dusting experiences as he worked his way through college in the mid-1950s. Robinet catalogued a long list of possible hazards: trees, livestock, kids throwing rocks, hunters shooting at the plane, bee stings, an engine quitting, crashing into a fellow duster, a bird hitting him, or the plane catching on fire. Lastly, he wrote, "you might succumb to one or more of the organic phosphates you were applying." In the summer of 1954, Robinet sprayed cotton with parathion dust in Mexico. "There was no such thing as a bad job with parathion," he explained, "because it killed horses, cattle, dogs, fowl, and pilots."[26]

Robinet lived the typical ag pilot life, which meant getting up early in the morning, working hard until the wind picked up, and then taking a break till afternoon, when the wind eased and permitted spraying. He usually drank hard with his buddies after work. Although he never flew in the South, Robinet's experiences with parathion no doubt were much the same as those of southern pilots. One of Robinet's friends, Jimmy Stenard, a Travel-Air pilot who crashed fatally in 1956, had never obtained a pilot's license. He flew in the wild and carefree days before World War II, and when the bureaucracy pressured him to get a license, he simply refused. Robinet respected, even admired, Stenard and was stunned when he learned of Stenard's fatal crash. "His airplane just rolled over on him during a turn back to the cotton field he was dusting. There was no other logical explanation or rationalization for the accident, other than the outside chance that parathion got to him, as it did to many others during those years," he wrote. Parathion, Robinet learned, "would impair your judgment and coordination to the point of affecting your turns and other functions requiring precise maneuvering and timing."[27]

Parathion was delivered to the airstrip as liquid in metal containers at 45 percent concentration, and loaders diluted it to 2 percent for application. "For our own safety, we were required to wear gloves and an organic respirator when flying," Robinet explained. Crews quickly learned how deadly the chemical was and how to deal with it. "A container of atropine tablets was taped to one of the structural members in the cockpit just in case the hopper was ruptured in a wreck and the pilot was splashed with the contents," he wrote. "The loaders were also required to keep a container of atropine tablets, along with an ample supply of soap and clean water, at the tank truck for emergencies." Robinet recalled an incident when an inexperienced loader spilled some of the concentrate on his foot, soaking his sock. When the pilot discovered what had happened, he had the boy remove his shoe and sock and wash his foot. "Fifteen minutes later the loader had passed out, and within two hours had quit breathing," Robinet related. The boy died a short time later. A lesser fate befell Buster Brown, who worked at the Mississippi Valley Aircraft Service, delivering chemicals to various locations and helping to mix them. On August 24, 1957, he splashed malathion into his right eye. The eye became inflamed, his sight blurred and then failed, and finally a doctor removed the eye.[28]

Pilots were required to get their cholinesterase level checked weekly. Unfortunately, the doctor who tested Robinet in 1956 was "a quack" who simply did a blood count. Before that year's spraying season ended, Robinet had close calls that he attributed to parathion poisoning. When his engine quit, he recalled, "I wasn't rattled but instead complacent; resigned to the fact that this was the fate that was in store for me." Another day he momentarily lost concentration and clipped a tree that he could easily have avoided. After that, he returned to the strip, went into the hanger, and sat on the floor. "I was terribly thirsty (one of the symptoms of parathion poisoning), but didn't have the strength to get up (another symptom)." Two weeks before the fall semester began, a friend flew him to Las Cruces to register for classes. "By this time, still without realizing it, the parathion was further taking its toll," he realized later. "My appetite was affected, I wanted to sleep but couldn't, I was feeling slightly dizzy and I was experiencing a headache. I rationalized the symptoms, attributing most of it to not enough sleep, too much booze and overwork." He went through registration "in a dream-like state." Back at the field, he worked two more weeks and continued his downward spiral. By the time he arrived for classes, he was extremely sick. "Everything I did was in a state of a drunken stupor. My nervous system seemed to be in shambles," he related. Then he visited the hospital and was given atropine. "I was ultimately treated and returned back into the jungle of academic activity, so blind that I could not read newsprint let alone text books. This condition lasted for only three days," he remembered. It took him a month though to regain his strength. Even climbing steps wore him out.[29]

In 1957 Robinet was again spraying parathion, and his doctor had learned how to test his cholinesterase level. Toward the end of the spraying season, he was loading his plane and let his attention wander, and "the end of the hose whipped out from the top of the hopper slinging goop all over the place; in the process completely saturating the front of my flying suit." He immediately washed off and then took off. As he later admitted, "What I should have done was to seek out medical help immediately instead of compounding my difficulties with additional exposure." His turns became ragged and his timing was off. "Pass after pass, load after load, things became increasingly foggy." At midday he went home for a nap but could not sleep; he then tried to eat but was not hungry. "I was definitely in no condition to fly but was afraid to admit it for fear of the usual sneers and snide remarks about being afraid of the airplane." Fortunately, the wind was too high to spray in the afternoon, and Robinet's brother, also his loader, fetched the doctor. "When they entered the room, all I could discern was distant and remote conversation. It was like all activity was taking place in another dimension. It was atropine time." He could not fly for three days. When he returned he was weak and simply drifted through the routine.[30] Robinet's experiences with parathion were frightening not just because of the symptoms that impaired his timing and endangered his life but also because of his macho attitude. Not even legitimate sickness was an excuse for failing to fly. Robinet survived, but given his harrowing account of parathion poisoning, there were probably other pilots who were not so fortunate.

Because tolerance for pesticides varied among people, reactions were unpredictable. Although some states required pesticide users to mark fields with warning signs, literacy, language, and laxity ensured that not every worker understood pesticide dangers. Robinet learned that some migratory cotton workers used empty parathion containers for seats in their houses and were killed. Another worker "merely hiked across a freshly treated cotton patch on the way to his shack and since he also didn't bathe, the parathion took its toll in short order. He went to bed that evening and was dead before the next morning." Parathion created collateral damage far beyond its insect target. "It killed everything, including rattlesnakes," Robinet recalled, "which were another hazard to the pickers."[31]

Robinet's experiences suggest a dangerous laxity among pilots, crews, landlords, and workers. Lethal pesticides become accepted as a normal means to kill insects. Given the dangers, it is remarkable that even more people were not poisoned fatally. Of course, in some cases the cause of death was not tied to toxic chemicals, and duster crashes were seldom investigated. An article that compared fatal duster pilot crashes with those of flying instructors revealed that between 1965 and 1979, 311 duster pilots died "from traumatic injury or immolation due to plane crashes" compared to 152 instructors.[32]

As Robinet's experiences suggest, many pesticide users had scanty knowledge of the toxicity of the chemicals they applied. In the fall and winter of 1969–70, the North Carolina State Board of Health investigated how farmers used, stored, and regarded pesticides. No doubt farmers' attitudes and practices had changed little over the decade. Investigators sampled 245 of Johnston County's 6,000 farms, where pesticides were used primarily on tobacco, cotton, corn, soybeans, and sweet potatoes. Most farmers stored chemicals within a hundred yards of their residences, and only 14 percent stored them under lock and key. Pesticide dealers and extension agents supposedly instructed farmers about pesticides. Nonetheless, nearly 70 percent of the farmers had never worn any safety equipment; 12 percent reported that over the past five years either they or a family member had been poisoned, and 8 percent knew of a neighbor who had been poisoned. While most farmers burned paper pesticide containers, many of them discarded empty bags, bottles, and barrels in trash piles in nearby woods. The report concluded that such practices created health hazards for farm families. It also faulted the extension service for not educating farmers about pesticide dangers. Yet city people also were careless. A parallel urban survey revealed a similar pattern of storing pesticides in kitchen cabinets, garages, basements, or bathrooms, not under lock and key. Of the 177 urban households polled, four knew of poisoning cases that had required a physician's care. Most people disposed of empty pesticide containers in their trash. The *New England Journal of Medicine* reported in 1967 that pesticides killed about 190 people a year.[33]

Chemicals used on crops or in the home were visible, and people who used precautions could handle them with little danger. Termite protection for new homes, however, was often invisible. Robert D. Dixon, vice president of Osmose, a wood-treatment business in Griffin, Georgia, admitted to self-interest when he complained in December 1969 of the excessive amount of chlorinated hydrocarbons used in termite treatment. Treated wood offered a benign solution to termites, he suggested. Dixon had worked in forestry before joining Osmose, where he had worked for nineteen years. One of the most common methods of termite control, he complained, was "soil poisoning," which used "enormous amounts" of chemicals applied underneath the foundation of new houses. He estimated that in 1968 alone, 40,000 Georgia houses were treated in this way, requiring a total of 4.1 million pounds of chlordane. Since 1955, he learned from the state entomologist, roughly 438,000 soil poisonings had been reported. The prescribed amount of chlordane per house was 103 pounds when treated according to Federal Housing Administration standards. Aldrin, dieldrin, and heptachlor guidelines called for 51.7 pounds per house, BHC guidelines suggested 82 pounds, and DDT guidelines specified 698 pounds. Dixon observed that many subdivision developers used soil poisoning, "so there are many areas where several hundred acres are treated, all going into the same watershed." He wondered if the USDA or the Federal Housing Administration knew "of the immense amount of chlorinated hydrocarbons going under housing." Soil poisoning was not the only method of termite control, he suggested, and he mentioned termite shields and treated lumber or even less toxic chemicals as an alternative.[34]

Termite treatment became big business both in new and in old homes. A number of incidents pointed to health problems when chemical fumes seeped into houses. Pesticides presented chronic and acute dangers, but they also could be misapplied, with dire results. In late 1956 or early 1957, a Missouri couple contracted with an exterminator to rid their house of insects. After the exterminator had sprayed a mixture containing malathion, the couple complained that the house "has been permeated in variant degrees with putrid, vile, toxic, noxious odors." They sued for negligence. On September 18, 1963, a Missouri landlord had three rental houses treated with chlordane while the residents were at work. They were not informed of the procedure. Upon returning to their house, Wilbur Hampton and his wife "noticed a strange odor." He turned on a bedroom exhaust fan, ate a piece of the apple pie that was sitting on a kitchen cabinet by an open window, and decided to take a nap. Within half an hour he was violently ill, and physicians concluded he was poisoned by chlordane residue on the pie.[35]

Gayle Laborde and her husband built a new house in Marksville, Louisiana, in 1978. She complained of sickness, consulted a doctor, and later sued five chemical companies that manufactured, distributed, or sold pesticides. In the trial, two physicians testified that she had been poisoned and that when she left the home for any time her condition improved. Twenty-two expert witnesses for the defense countered those claims. Wayland J. Hayes Jr. testified that the pesticide values in Gayle Laborde's blood were unexceptional and that she was not suffering from exposure to chlordane, heptachlor, or organophosphates. Laborde lost her case. In Georgia, a claimant for workman's compensation stated that he had been poisoned by parathion on a Saturday. He experienced nausea, vomiting, and diarrhea, and he ran a fever. On Monday he became mentally disturbed; several days later he was committed to a mental hospital. Expert witnesses disagreed about whether or not parathion could produce those symptoms.[36] There was a prevailing assumption that experiments on rats or mice would correlate with chemical effects on people, yet people often reacted to pesticides quite differently as a result of allergies or chemical peculiarities. Although physicians could not agree on whether the illness of Gayle Laborde or that of the claimant in the workman's compensation case came from chemical exposure, their cases resembled that of Charles Lawler in suggesting unique human reactions to chemicals. Chlordane applied under houses invaded indoor air and continually exposed residents to toxins. While some people could withstand the fumes, others developed respiratory and neurological problems.

In June 1987, congressional hearings on a bill to ban chlordane from being used in home termite treatment aired disturbing problems. When Congressman Tom DeLay complained that the hearing was "weighted toward those that want to ban chlordane," Congressman Henry Waxman reminded him that representatives from Dow, Shell, Terminex, and other companies had not accepted invitations to testify. Drawing on hackneyed and dubious analogies, DeLay compared the bill that would ban chlordane to one that would "outlaw automobiles" and later reminded the committee that "aspirin is a health hazard when misapplied." Using National Academy of Sciences statistics, Waxman calculated that three people out of a thousand in homes treated with chlordane were at risk for developing cancer. DeLay minimized the estimate, saying, "That's 0.3 percent." Reminding DeLay that 100 million people had been exposed to chlordane, Waxman replied that 300,000 people were at risk. DeLay labeled the statement "inflammatory" and strongly defended the continued use of chlordane.[37]

Witnesses before the committee raised grave questions about chlordane termite treatment. On July 27, 1983, Michelle Slowey of Portsmouth, Virginia, had her house treated for termites with a compound called Termide, a mixture of chlordane and heptachlor. The Slowey family complained first of a noxious smell, several days later of skin rashes, and "after 2 or 3 weeks, loss of memory, nausea, vomiting, diarrhea, tremors, convulsions, ringing in the ears, bleeding from the bladder, loss of energy, drowsiness, skin lesions, and wart-like lumps that were on our skin." Michelle Slowey had called the exterminator a few days after treatment to complain that the chemicals had seeped into the house, destroying the parquet floor and wall-to-wall carpeting. Her calls to local, state, and national health departments elicited no response. Finally the USDA sent a field representative to her home on October 3. Six weeks later she learned that "25,000 micrograms of chlordane was detected in all the solid samples, and 700 micrograms of heptachlor was detected in the solid samples that they took from my home." On November 4, the USDA judged that although "an excessive emulsion was applied . . . we are unable to document a misuse." It was February 1984, Michelle Slowey testified, before an official at the Environmental Protection Agency (EPA) notified her that her house "had to be destroyed, and he told me to call Velsicol Chemical Corporation." In April she attempted to buy another house, but insurance problems forced the family back to the contaminated house. She then gave up her children to her former husband and moved out. It was 1985 before she secured proper medical attention for the family through an eighteen-week "lymphophalmic absorbent treatment of chlordane and neurotoxicity" program at the University of Virginia. Over time, she recalled, her brass lamps "exploded and disintegrated," the chemicals ate through silver plate to its copper base, and her furniture and clothing "began to disintegrate." Michelle Slowey had amassed four binders with documentation of the problems resulting from the calamitous termite treatment, and she had bills for $64,000 in medical costs. At the time of the congressional hearing, she had been unsuccessful in collecting damages. She found no comfort in the estimate that three out of a thousand residents in homes treated with chlordane would develop cancer.[38] It seemed a high price to pay for termite control.

Juanita Russell, a Delta Cultural Center volunteer, in Helena, Arkansas, recalled the time twenty years earlier when her house was treated for termites. The man who applied the pesticide could not read the directions and applied chlordane under the wool rugs inside the house. "The next day," she said, "we were in intensive care in Memphis, Tennessee, Baptist Hospital." When she complained to her landlord, he had the air conditioner repairman check the house, who recommended simply changing the filters. Juanita Russell moved, but she still has lung problems and suffers from emphysema and chronic bronchitis. No one could ever live in the house, and it had to be destroyed. "My bible today still smells of the chlordane," she lamented.[39] The EPA finally banned chlordane in 1988.

As chlorinated hydrocarbons were being phased out in the early 1970s, eastern North Carolina tobacco farmers turned to organophosphates. Many farmers did not at first grasp the contrasting actions of the two chemical families. They were accustomed to DDT, which presented no immediate danger to users. Parathion, which replaced DDT as the insecticide of choice for hornworms, was extremely dangerous when applied but dissipated after several days. Clarence Lee Boyette, who farmed near Pink Hill, North Carolina, used parathion—locally nicknamed "Big Bad John"—to kill his hornworms. He poisoned his tobacco fields and waited more than the recommended five days before reentering them. His seven-year-old son Daniel was in the field with the crew on July 31, 1970. In the middle of the night Daniel complained of a chill and asked for covers; the next morning he was dead. Five days later his brother Curtis fainted at the barn; he was rushed to a hospital, given atropine, and survived. The whole family had such high exposure levels that doctors speculated one more day barning tobacco could have sickened them all.

In a similar incident, in early August 1970, Jay Adams, a seventeen-year-old who lived near Sanford, North Carolina, became ill when working in his uncle's tobacco field a day after it had been sprayed with parathion. Two weeks later he went back to the field, became sick again, and died two days later. "It's advisable not to ever go back again," Dr. Shirley K. Osterhout at Duke's Poison Center warned. "We've noticed a lot of cases where the victim just piles up daily doses. He seems to be all right, but suddenly he passes out, goes into convulsions or becomes paralyzed." Parathion poisoning also sickened fruit and vegetable workers in California and Florida. Two workers died in Dade County, Florida, early in 1970. Reacting to these incidents, Dr. Martin P. Hines, director of the North Carolina Board of Health, observed that "parathion is too deadly a pesticide to be distributed without any type of control on it." It was in many households and on farms and not under lock and key, he protested, declaring, "Something should be done at a national level to restrict the use of such a potent and lethal pesticide." On November 30, 1970, the USDA announced safety precautions for parathion, and Ned Bayley recommended new labeling with the "wholehearted support" of the pesticide industry.[40]

The parathion poisonings prompted the USDA to investigate, but chemical companies argued that organophosphate poisonings showed the wisdom of reinstating less lethal chlorinated hydrocarbons. In North Carolina there were an estimated thirty parathion poisoning cases during the 1970 tobacco barning season. Since January 1967, the USDA admitted, it had investigated twenty-three accidental parathion poisoning incidents that had killed 132 people.[41] In the summer of 1971, former PRD head Harry Hays, accompanied by James O. Lee (who worked for the newly created EPA), investigated the use of parathion among tobacco farmers in Quincy, Florida, and Raleigh, North Carolina. Florida shade-grown tobacco provided cigar wrappers and thus relied upon high-quality leaves that showed no insect damage. "The most commonly used pesticides," Hays discovered, were "diazinon, Di-Syston, Dylex, endosulfan, Guthion, Lannate, malathion, and parathion." Human studies of parathion's effects on shade tobacco workers had not been completed, but Hays concluded that it "has caused no serious difficulty."[42]

At North Carolina State University, Hays talked with scientists studying parathion's effect on Florida tobacco workers. During the 1971 growing season, scientists followed forty-seven volunteer tobacco workers. They traced cholinesterase levels for ten days following application of parathion and found levels "within the normal range and thus . . . of no toxicological significance." Parathion residues on tobacco leaves were from 0.05 to 4.8 ppm during harvest. Hays decided that additional research on human health was not needed in North Carolina despite the vast difference in the work culture of flue-cured tobacco.[43]

During each week of a six-week harvest season, flue-cured tobacco primers moved through the fields taking three or four leaves from each stalk. They usually began work at daybreak in dewy fields. The damp leaves were handled by children and women at the barn as they were bundled and tied onto sticks. Tobacco tar and presumably pesticide residues stuck to any exposed part of the body. Hays's assumption that data gathered from Florida shade tobacco workers also could be applied to flue-cured hands demonstrated a startling misunderstanding of the flue-cured work routine. Flue-cured tobacco farmers used chemicals for aphids, flea beetles, budworms, and hornworms, and many used parathion two or three times a year. Hays admitted that chemicals could have affected some North Carolina tobacco workers, but he dismissed parathion as responsible for the five deaths and fifteen sickened people in the summer of 1970. Instead, he asserted that "the fatalities involved were either deliberate misuse or unavoidable accidents." He judged that twelve of the fifteen illnesses were treated as parathion poisoning "on the basis of signs rather than cholinesterase determinations." Medical records were incomplete, he claimed. After observing that tobacco workers were in the sun all day and could become dehydrated, he added, "It is more likely that these reported illnesses were due to dehydration rather than to parathion." Hays recommend a five-day period after parathion spraying before workers entered the fields. He rejected a recommendation that "placards be placed around a field where parathion has been used," arguing, on the one hand, that workers would not pay attention to the signs, and, on the other, that a worker "could use it to his own advantage." Hays's attitude summed up ARS callousness toward workers: they were dumb and lazy. When Hays presented his sanitized report before representatives from Monsanto, Stauffer, the federal extension service, and the EPA, some of them expressed concern. "Indeed," reported ARS deputy Waldemar Klassen, "after the meeting, a representative from EPA told me that he personally feels that the registrations for parathion use should be cancelled."[44]

In September 1971, the ARS recommended that tobacco farmers seek a less toxic poison than parathion, or, if they continued to use the substance, to delay entry into their fields. An unsent draft of the recommendation observed that "only properly trained personnel should apply parathion and supervise work in treated fields." The draft queried the role of industry and the ARS in "ensuring the availability of safe and effective agricultural chemicals of all kinds." A task force was studying alternatives to "parathion and other hazardous chemicals." The draft also raised disturbing questions about other chemicals used on tobacco. Although MH-30 (maleic hydrazide) was "considered to be safe," Penar "causes severe dermatitis." Workers also reported experiencing nausea, which could come either from pesticides or toxins within the tobacco plant. The draft letter recommended further study to investigate "heat, dehydration, physical exertion, and chemical treatments." These statements were eliminated before the final recommendations were circulated. The final letter detailing parathion safety expressed complete confidence in Harry Hays's report and declared that the only practical means of avoiding parathion poisoning "is to delay reentry into fields until residues have dissipated to toxicologically insignificant levels."[45]

When in the early 1960s concern increased about the relation between smoking tobacco and lung cancer, the ARS had expressed concern about pesticide residues on tobacco. Tobacco was grown in fields and contained residue from insects and wildlife. It was handled by hand at harvest with little thought about sanitation. As mules plowed or pulled tobacco slides through the fields, they often splattered the leaves with excrement. Tobacco stored in packhouses accumulated mice excrement, dead spiders, and other insect carcasses. In factories, workers picked up cigarettes by hand and placed them in trays to await packing. From the plantbed to the cigarette, there was no concern for sanitation.

Yet the natural ingredients that ended up in cigarettes were not as worrisome as invisible pollutants. In August 1971, C. H. Hoffmann, acting director of the ARS's Entomology Research Division, produced a long-overdue research agenda for tobacco pesticides. Despite opposition from cigarette manufacturers and foreign buyers, Hoffmann discovered, farmers applied MH-30 before harvest for sucker control. Although not as toxic as organophosphates, MH-30 could cause anything from skin, eye, nose, and throat irritation to convulsions and coma. He revealed that Penar, also used for sucker control, "causes severe dermatitis and is becoming unpopular for this reason." If it turned out that Penar caused pulmonary adenomas, George W. Irving Jr. had mused in July 1970, "we would be hard pressed to explain why we allowed it to be used while the studies were under way." Nicotine and other toxins in tobacco could also be harmful, the report speculated. "Some workers have reported nausea, particularly on rainy days when the plants are wet."[46] The report showed a startling lack of research on pesticides that were applied to a product that was ultimately inhaled into the lungs.

While manufacturers complained that MH-30 affected cigarette aroma, European buyers threatened to shun chemically treated tobacco for health reasons. When in 1967 Germany ruled that tobacco was a food crop, the ARS was caught short with no residue studies. "The increased apprehension of the tobacco industry over the public concern about the reputed hazards of insecticide residues in cigarettes and other manufactured tobacco products," an ARS study had warned in 1960, "demands that greater emphasis be given to development of methods and materials for insect control that will reduce or avoid residue problems." A decade later the ARS was still defending pesticides. It was not just health concerns that spurred the ARS to look into Penar and MH-30 toxicity. In October 1971, ARS science and education deputy Sam R. Hoover evaluated tobacco issues that would come before the November meeting of the Southern Science and Education Regional Workshop in Mobile, Alabama. "The Germans have announced intentions to outlaw the importation of tobacco treated with MH-30 after 1975," he revealed. "The British have ruled against Penar because of tests done in Great Britain which are fairly conclusive in showing that Penar is carcinogenic at relatively high levels." The German government did not allow its farmers to use MH-30, and Canada had found substitute chemicals and had not used either MH-30 or Penar for two years. "However, some American workers, most notably Dr. Ken Keller of North Carolina State," Hoover added, "consider that the use of MH-30 is almost mandatory, especially for mechanically harvesting tobacco."[47] As other governments attempted to protect their citizens' health, both ARS bureaucrats and U.S. agribusiness interests dismissed public health issues.

As more questions arose about residues on tobacco leaves, the ARS admitted serious questions about the problem. "We have concern for the health of persons manufacturing the product and for those applying the pesticide chemical to tobacco plants, and to the persons using the tobacco products," Harry Hays admitted to Senator Hiram Fong in November 1967. The ARS conducted research on residues, he claimed. While the FDA set tolerances on raw agricultural commodities, the USDA established residue limits on tobacco. Hays did not elaborate.[48]

Other food products moved to market with little regard for toxicity. Cotton lint was spun to make clothing, and cottonseed oil was marketed as vegetable oil. Like tobacco, cotton and cottonseed oil had not been targeted for pesticide tolerances. When L-10, a defoliant, was applied to cotton on October 29, 1965, before its harvest on November 1, the procedure "resulted in arsenic acid residues in the lint." A Clemson University researcher who handled the samples "developed skin irritations." Analysis revealed a range of 0 to 126 ppm arsenic in the lint. No tests were done on cottonseed that, when crushed, yielded vegetable oil. Given the interval between spraying L-10 and harvest, the report concluded, "high levels of arsenic would be expected to occur in the lint." If the residue levels were present in clothing, the report hinted, "serious problems might occur." There is no indication that the ARS followed up on this report.[49]

In 1967, H. Wayne Bitting, a staff specialist in the ARS's product evaluation section, submitted a research proposal on cyclopropene acid residue in cottonseed oil. "Research results with laying hens fed cyclopropene acids," he wrote, "showed a high embryo mortality and rat experiments showed that these acids inhibited the performance of some body enzymes." If the FDA set no tolerance level, he concluded, then cottonseed oil "would be relegated to industrial uses." Thus the FDA allowed the interstate shipment of adulterated vegetable oils. "At present," an ARS administrator admitted in May 1967, "it appears that the Food and Drug Administration has no concern, from a human health standpoint, with chlorinated hydrocarbon pesticide residues in edible vegetable oils." The residues came primarily from crop rotation, when persistent pesticides invaded the next crop. Legally, soybean growers were in a risky position, for contaminated beans were shipped in interstate commerce. "If FDA were to enforce present regulations," the ARS concluded, "there would be practically no Southern soybeans processed."[50]

While the ARS continued its dalliance with industry supporters, increasing pressure prompted a thirty-day ban on the use of chlorinated hydrocarbons in state-federal cooperative programs. In response, Melvin C. Tucker, chairman of the Southern Plant Board, complained to the ARS's J. Phil Campbell Jr. His July 1969 letter epitomized zealously propesticide arguments that were at that moment justifying another fire ant campaign. Members of the Southern Plant Board were "shocked and alarmed," Tucker wrote, because the ban "is being editorialized as a prized victory by the anti-insecticiders and a giant step toward their goal of total banishment of these materials for all agricultural uses." Tucker reminded Campbell that large segments of the agricultural community were "entirely dependent" on chlorinated hydrocarbons. "Sudden discontinuance of their use would have much the same effect upon the security of our nation," he thundered, "as the withdrawal of nuclear weaponry from the armed forces." Tucker suggested that the USDA mount a counterattack. "Added to this would be all of the powers, resources and able voices of the pesticide industry, the backing of such organizations as the National Agricultural Chemicals Association, and the support and help of all the states," he argued. Tucker faulted the USDA for caving in to "pressure" that was not based on scientific studies or "a crisis or emergency." The Southern Plant Board, he continued, urged the USDA to "protect agriculture, not only from plant and animal pests but from human pests as well." The anti-insecticiders "pose more of a threat to our economy than many of the invertebrates we are fighting." In his enthusiasm, Tucker lapsed into insect-control terminology, writing, "These antis can be eradicated, or at least controlled, by decisive, vigorous leadership from the Department." The USDA should combat anti-insecticiders' "erroneous statements and scare tactics" with "truthful factual, documented evidence" and, he added, with good public relations. Dave L. Pearch, Louisiana Commissioner of Agriculture, wrote a slightly less strident letter urging Campbell to make people aware of the positive benefits of chlorinated hydrocarbons.[51]

Chemical companies and their allies could not stop the drive to ban DDT and other chlorinated hydrocarbons. In November 1969, the *Federal Register* carried a notice of intent to cancel nearly all uses of DDT, even though farmers, grower associations, and local and state officials complained that there were no adequate substitutes. By this time, scientists had associated chlorinated hydrocarbons with thin eggshells and declining populations of predator birds, such as eagles, hawks, and pelicans. After the *Federal Register* notice had been published in local newspapers and had gained a wider distribution, the PRD was flooded with anti-DDT letters. "The letters came from all parts of the country," an alarmed George W. Irving Jr. exclaimed, "from grade school children, high school and college students, graduate students, teachers, professors of law, sociology and medicine, housewives, civic leaders, farmers, conservationists, and members of nature clubs." Writers condemned the USDA for not banning DDT outright, deplored environmental pollution, and claimed that birds were disappearing. "The impression given," Irving concluded, "was one of mass hysteria."[52] The response against DDT, of course, was a tribute to Rachel Carson's warning in *Silent Spring*. Chemical proponents had branded Carson a hysterical woman in their attempt to undermine the authority of her argument. In fact, chemical supporters were projecting their own hysterics. Ultimately, Carson and her disciples dried up the production and use of most chlorinated hydrocarbons. Because of their persistence, however, these chemicals linger in the environment.

The ARS also maintained contact with scientists who were eager to discredit claims that chemicals were dangerous. Robert White-Stevens, who had attacked *Silent Spring* and had generally defended DDT use, continued his crusade. In December 1970, he requested that Ned Bayley send him the bibliography that accompanied the USDA's brief in a suit brought by the Environmental Defense Fund. White-Stevens was preparing a lecture for Rutgers University defending DDT and other chlorinated hydrocarbons. He expected it to be well attended, and, he boasted, it "may develop into a considerable 'donnybrook.'" He was planning to discredit the claim that all DDT ever manufactured was still active in the environment. "I am considering compiling an annotated pamphlet or book listing on the left hand (appropriately) pages the anti-DDT claims and on the right the scientific evidence that refutes them. The problem will be, of course, that for every hysterical pseudoscientific claim such as the one mentioned above there are at least 10 publications that refute it." Bayley sent the bibliography. "We believe a publication of this nature would be extremely valuable and sincerely hope that you are able to complete this project," Bayley wrote encouragingly.[53]

Despite the work of pesticide advocates and the continuing growth of pesticide use, gradually the environmental wind had shifted, and toxic drift provoked widespread distrust of pesticides. In 1970 there were four lawsuits pending against USDA pesticide projects. One involved mercury poisoning in New Mexico, the second was an Environmental Defense Fund suit to ban DDT, a third challenged the mirex fire ant control program, and the fourth sought to suspend 2,4,5-T use on food crops. In addition, there were thirty-nine "requests for advisory committees or public hearings pending in the Department concerning refusals of registrations of economic poisons, and cancellations and suspensions of registrations."[54]

Increasing awareness of wildlife kills, poisoned domestic animals, and threats to human health ultimately overwhelmed the ARS and propesticide defenses. Committee hearings, continuing publicity about pesticide dangers, and highly publicized accounts of pesticide accidents undermined chemical advocates. In November 1969, Dr. Emil Mrak's report of the Commission on Pesticides and Their Relationships to Environmental

Health suggested a shift from persistent pesticides and called for a single agency to regulate pesticides. Department of Health, Education, and Welfare (HEW) Secretary Robert Finch announced a phase out of all but essential DDT uses. The creation of the EPA in December 1970 combined ARS registration, HEW tolerances, and the Department of Interior's research. In theory, its goal was to protect the environment and not corporate interests, but, ominously, Congressman Jamie Whitten maintained control of EPA appropriations.[55] The first Earth Day took place April 22, 1970, and the public looked at the momentous occasion as the beginning of the end of the environmental crisis.

7
HAZARDOUS LABELS

How smoothly one becomes, not a cheat, exactly,
not really a liar, just a man who'll say anything for pay.

Tom Rath in Sloan Wilson's *The Man in the Gray Flannel Suit*

For what is a man profited, if he shall gain the whole world, and lose his own soul? Or what shall a man give in exchange for his soul?

Matthew 16:26

After World War II, synthetic pesticides stampeded across the land with scant toxicity research and shaky directions for proper use. The ARS's Pesticide Regulation Division (PRD), tucked into the mammoth USDA bureaucracy and shielded by ARS leadership, became the nerve center of the country's pesticide registration policy. Despite its crucial public health mission, the PRD operated behind closed doors, where a chorus of eager corporate registrants drowned out concern for pesticide safety. Malfeasance and perfidy allowed a percentage of the more than 45,000 products containing 900 different chemical compounds to gain access to and remain on the market mislabeled, not fully tested, and dangerous. Had chemical companies written the legislation, staffed the PRD, and dictated crucial decisions, their influence on research, regulation, and label policy would not have been more absolute. The general public simply was not aware of the labyrinthian and flawed process that supposedly protected it.

The registration process written into the Federal Insecticide, Fungicide, and Rodenticide Act of 1947 stipulated that a manufacturer furnish the name of the chemical, the composition of the pesticide, safety reports, residue tests, analytical methodology, and a proposed label. The PRD and other relevant agencies evaluated this information in the context of possible threats to fish, wildlife, and humans and labeled the product accordingly. If a chemical was to be used on food, the Food and Drug Administration (FDA), part of the Department of Health, Education, and Welfare (HEW), examined the manufacturer's data, analyzed the product, and set a tolerance, that is, how much residue could safely remain on food when it went to market. Without a tolerance, foods with residues were considered adulterated. By July 1963, there were 2,500 tolerances for approximately 130 chemicals being used on raw agricultural products. The Public Health Service (PHS), also under HEW, reviewed the data for health issues. The Department of Interior's Fish and Wildlife Service (FWS) commented on possible dangers to fish and wildlife.[1] Although the approval process was spread over a wide bureaucratic domain, the PRD hoarded final decision making.

Despite its distrustful relationship with other government agencies and its pliant association with chemical companies, the ARS insisted that it carried out its registration and enforcement duties. In 1963, the ARS's M. R. Clarkson testified before a Senate committee that registration was a rigid scientific process. The tens of thousands of formulations, he stressed, "include a great variety of toxicants." In their registration requests, manufacturers submitted "detailed and convincing test results" on the chemicals' effectiveness, as well as "extensive toxicological studies." ARS scientists tested the chemicals on small and sometimes large animals and often turned to experts outside the agency. It was "routine procedure," Clarkson claimed, that the ARS consulted with the FDA about "uses of pesticides on foods," enforced a "stringent" label policy, and accepted a label "only when they are satisfied that the directions and warnings it gives are adequate to protect the public." The only flaw in the process, Clarkson admitted, was that corporations could register formulations "under protest" and continue to sell the product until the USDA developed "legally acceptable proof to justify a court decision to remove it from the market." He failed to mention that the ARS had never taken such action.[2]

The ARS maintained five analytical laboratories that tested samples before registration and six laboratories in entomology, plant biology, bacteriology, pharmacology, and toxicology that tested marketed products. PRD staff also inspected and sampled pesticides on the shelf and in warehouses, checking the accuracy of labels and contents. Because the ARS stubbornly refused to cooperate with other agencies, a 1964 interdepartmental agreement directed it to work closely with the PHS, the FDA, and the FWS. The ARS ignored the agreement. In the first five years after protest registration ended in 1964, the PHS raised 1,633 objections to proposed registrations. Instead of submitting contested cases to the Secretary of Agriculture for resolution as stipulated in the agreement, however, the ARS/PRD unilaterally approved the applications. Nor did the ARS convene annual meetings to reconcile problems with other agencies, as the law required. Ignoring input from other agencies freed the ARS to rubber-stamp all registration applications.[3]

Clarkson's assurances aside, the ARS continually stalled the PHS and only reluctantly shared residue data. Thomas H. Harris, chief of the division of pesticide registration at the PHS, complained on January 19, 1967, that he lacked the data needed to "complete our pesticide label review in the time specified under the Interdepartmental Agreement." On June 19, 1967, Harris suggested that each potential registrant supply toxicological data to both the PRD and the PHS. "There are numerous pesticide chemicals for which we have requested data but such data have not been received," he complained to PRD director Harry Hays. "In some instances we have been informed that no data are available," he added, raising questions of the PRD's methodology and cooperation. Even though many chemical formulations had been registered years earlier and "have a history of apparent safe use," Harris continued, "we believe that at least some toxicologic data should be available." He offered to furnish a list of such products.[4] Without such data, the PHS could not evaluate a formulation's toxicity as required by law.

Harry W. Hays became chief of the PRD in 1966. Receiving his Ph.D. degree from Princeton University in 1938, he joined Ciba Pharmaceutical Company as assistant director of research. After teaching at Wayne State University College of Medicine for nine years, he became director of the Advisory Center on Toxicology for the National Academy of Sciences National Research Council, where he remained until he joined the PRD. In his reply to Harris, Hays insisted that by law all registration information went only to the PRD. He offered to forward data "after it has been reviewed by our staff." The PRD relied upon both published reports and its files, Hays explained, but it was sometimes difficult "to connect the information with a given product." He added that the PRD was in the process of organizing its files. "When we have finally assembled all of the data that has accumulated over the years," he hedged, "we hope you and your associates will feel free to use the reference library at any time." Hays's draft reply to Harris elicited marginal comments. One ARS administrator scribbled, "Do you think there's any danger that maintaining a hard line with PHS will breed new problems?" Another staffer suggested that if difficulties arose, "we can change."[5] In essence, Hays's reply was calculated noncooperation. If the files were as disorganized as Hays suggested, the PRD surely had difficulties making informed decisions on label applications. The refusal to share basic information with the PHS suggests a propriety mentality, one that enshrouded pesticide registration in bureaucratic rhetoric and that neglected scientific data and experts from other agencies.

Thomas H. Harris deplored the PRD's noncooperation. "We don't see the final, printed label," he complained. When Harris requested information from the PRD on toxic chemical registration, he received no reply. He phoned Harry Hays on April 2, 1969, explaining that he needed the information to determine how many products had been registered over FDA objections. Hays at first insisted that such information was unnecessary, but Harris read to him from the agreement that required the USDA to furnish him registration information. According to Harris's phone log, Hays admitted that after a series of Government Accounting Office (GAO) investigations in 1968, which had been highly critical of the PRD, "he was suspicious of everybody and that he thought the GAO was out to get him." When Harris pressed him for the information, Hays refused to set a timetable.[6] As Hays feared, the critical 1969 GAO reports prompted L. H. Fountain to launch an investigation that led to hearings in May and June 1969.

The PRD employed the same obstinance with the FDA. Reo E. Duggan, the FDA's deputy associate commissioner for compliance, testified about the cooperative agreement before Congressman L. H. Fountain's subcommittee on June 24, 1969. Duggan countered the ARS's persistent call for additional research before it would cancel a label by arguing that "the lack of scientific evidence is sometimes as important for consideration as positive evidence." He used the example of thallium-based rat poisons. "Place some of the substances in any household where a child can get hold of them, and they do get them," he pointed out. "What kind of scientific evidence can you give to support that sort of position?" Each week the FDA forwarded to the ARS its objections to proposed registrations. From July 1, 1968, to June 1, 1969, for example, the FDA forwarded 177 objections. The ARS neither informed Duggan how many formulations were registered over FDA objections nor kept a statistical log of PHS objections. As he had done with Thomas Harris, Harry Hays refused to meet with Duggan. On June 24, 1969, Congressman L. H. Fountain asked how long the FDA and the PRD had been having meetings to discuss their disagreements on label warnings, Duggan replied, "The first one is scheduled for tomorrow."[7]

In the 1969 Fountain hearings, Harry Hays attempted to put the best light on PRD cooperation. New Jersey congresswoman Florence P. Dwyer asked Hays how many products were marketed that the PHS had "expressed doubt concerning their safety or efficacy." Hays replied that out of the 45,000 products registered by the PRD, "I would say a half dozen in which the Public Health Service has expressed some concern." Dwyer asked for the record, which revealed that out of the 11,361 product labels reviewed by the PHS between January 1, 1968, and March 31, 1969, there were 252 unreconciled objections. Either Hays had little grasp of his agency's work or he purposely misled the committee. Ultimately, when the duties of the PRD were transferred to the Environmental Protection Agency (EPA), other agencies were consulted as the agreements directed. Under the George W. Bush administration, however, a rule was made that allowed the EPA to evade input from wildlife agencies, signaling a return to the rogue days when the ARS handled pesticide issues.[8]

While the PRD stonewalled other government agencies, it encouraged corporate representatives to discuss their registration problems. Even as chemical companies feigned cooperation with registration rules, they ruthlessly exploited pliant PRD bureaucrats and probed for loopholes in registration requirements. When faced with registration problems, claiming the need for more research became their mantra. Three years after the PHS report on Mississippi fish kills, the ARS bureaucracy was still debating whether or not to take endrin off the market. When gas chromatography analysis revealed formerly undetected endrin residues in crops, chemical companies scrambled to sanitize their data. Since 1965 Shell Chemical Company and the FDA had negotiated residue levels; whenever FDA decisions did not please Shell, the company threatened the long, drawn-out process of appealing to an advisory committee. As pressure built to curb endrin in the spring of 1967, Shell and Velsicol Chemical Corporation notified the ARS and the FDA that six complex studies were underway that would justify preserving endrin registrations. Meanwhile, the corporations volunteered to abandon some product registrations "which might cause regulatory problems disproportionate to their beneficial contribution." Eager to coordinate the initiative, ARS associate director Robert J. Anderson suggested that Shell and Velsicol meet with the PRD "to determine what endrin uses could be voluntarily withdrawn before setting up a meeting with FDA." After agreeing on tactics with the ARS, corporate representatives submitted a joint letter to the ARS and the FDA reiterating that important endrin residue studies were underway. "Endrin," the letter stressed, "has been extensively, beneficially and safely used in the United States and many other countries for more than a decade."[9]

Despite the careful groundwork, at a May 10, 1967, meeting, FDA representatives did not share the ARS's enthusiasm for preserving endrin registrations. The FDA's Reo Duggan "expressed some skepticism as to the practicality of any level of endrin which could be permitted in food." Another FDA staffer "expressed the feeling that endrin should be completely eliminated." Harry Hays, as previously agreed, promised to ask Velsicol and Shell "to agree to the cancellation of any uses which are not of particular importance to them." The meeting broke up without making a decision about endrin.[10]

Two days later, ARS officials met with Shell and Velsicol representatives. As agreed previously, the corporations "submitted a list of uses which they would like to continue and which, in their opinion, would not create a serious residue problem." In a June 8 meeting with the ARS and the FDA, Shell's Bernard Lorant, vice president for research and development, formally requested extensions of the registrations for several years until continuing research projects were completed. Harry Hays was keen to cooperate, insisting that residue levels "do not, in my opinion, constitute a significant hazard." On August 4, the FDA's advisory committee on aldrin and dieldrin gave interim approval of existing tolerances.[11] At every step of the review process, ARS staffers worked closely with Shell and Velsicol. By yielding on several registrations and beginning new studies, corporations kept endrin on the market. Such tactics worked until endrin was banned in 1986.

Even such close cooperation between chemical companies and the ARS could not silence growing environmental concerns among the general population. Because of improved detection technology, chlorinated hydrocarbons such as DDT, aldrin, dieldrin, and endrin came under increasing scrutiny. Even as the PRD worked with chemical companies to meet new standards, friction occasionally erupted. On August 27, 1967, Shell general manager K. R. Fitzsimmons upbraided ARS administrator George Irving Jr. for "very disturbing" developments. "The implication is that, regardless of all the expert committee studies, the status of experimental effort and in the absence of all the facts, persistent pesticides are to be eliminated," he fumed. Given that Shell had complied "in every detail with all statutes, implementation orders, and expert committee recommendations" and "the fact that there is no evidence indicating any hazards to the public health," Fitzsimmons continued, "I cannot understand how your staff can continue to threaten the removal of our label." He guilefully suggested a breakdown in communication. "As responsible members of the agricultural community," he lectured Irving, "we cannot and will not accept complacently decisions being made in the absence of overriding scientific facts." Fitzsimmons snidely asked Irving's opinion as to what Shell might do to "prevent further erosion of our business until all data is gathered, put into the risk vs. benefit equation, and properly integrated." On September 7, the ARS succumbed to Fitzsimmons's bullying rhetoric and agreed with Shell on interim tolerances.[12] Instead of halting chlorinated hydrocarbon use until Shell's studies were completed, the ARS allowed possibly dangerous compounds to remain on the market while the studies were in progress.

In 1967 the Federal Committee on Pest Control questioned whether labels were adequate to protect the public. Harry Hays suggested that the PRD lacked "scientific knowledge" on the subject. He admitted that a recent two-year University of Wisconsin study "indicated that a fairly high percentage of pesticide users do not really understand terminology commonly used in referring to pesticides." Hays mulled over questions about readability, technical terminology, color schemes, print size, and format. By even entertaining such questions—small print, excessive verbiage, poor design, vexing color schemes—he conceded that labels were inadequate and confusing. "What can be done," he nevertheless fretted, "to induce a pesticide user to read, understand, and follow instructions on the labels designed to protect him?" With a supreme grasp of bureaucratic delay tactics, Hays recommended that the PRD contract with another university to review label literature and then hire "an advertising agency" to design sample labels. When these steps were completed, "we would begin working with the chemical companies and our own label review staff to make pesticide labels as effective as possible." In 1969 the PRD contracted with the University of Illinois for such a label study, and Hays expected the results within a year. He did not express urgency or begin revisions of labels using common sense and PRD staff.[13]

Hays's dilatory plan made sense only if the Wisconsin study was inadequate, the proposed Illinois study necessary, and the PRD label staff incompetent. The process consumed time and awarded contracts but failed to address pesticide label shortcomings. By the summer of 1967, synthetic pesticides had been on the market for twenty-two years, but the PRD seemed content to allow faulty labels to persist for another decade. Hays did not even address such significant label problems as inaccurate statements or the lack of vital information. The PRD possessed enormous power in designing and approving labels for pesticides that, used as directed, led to serious—or even fatal—consequences.

Even when the PRD cancelled dangerous labels, it failed to recall or seize products. Rat poison endangered children who ate the thallium paste spread as bait on bread or cheese. In 1960 the PRD revised labels to cut back on the amount of thallium in the poison. Still, the PHS estimated that roughly four hundred thallium poisoning accidents took place in 1962 and 1963. At L. H. Fountain's 1969 congressional hearings, committee counsel James S. Naughton suggested that the thallium label was contradictory when it cautioned users to keep it from children. "If you keep it away from children it is not going to kill the ants and the rats, isn't that true?" he asked sarcastically. He derided Harry Hays's protracted label study. "Of course, you don't need a contract with the University of Illinois and a 2-year study period to know that if you put rat poison on the floor the children are going to get at it," Naughton chided. "In 1965, after who knows how many more deaths had resulted, the registration was finally canceled," he continued, "but the GAO was able, in 20 percent of the establishments in the Washington area which they sampled, to buy thallium products in January 1968." As late as 2004, rat poison formulators resisted regulations to add child warnings to their labels. In 2003, fifteen thousand children under six years of age "ingested rat poison." To its credit, Naughton admitted, the PRD removed "attractive nuisances," highly toxic poison packaged in "little cartons in the shape of a garage or a church with a steeple." But even when the PRD discovered a product that violated registration rules, it seized only the inventory in the store where it was found. It seldom examined manufacturer's records to locate other lots. Only twice had the PRD used its seizure powers to clear products from the market. In 1962 it seized interstate shipments of "Steri-Fleece," which, when used on diapers of premature infants, caused methemoglobinemia; in 1963 it seized a Rotenone product that was contaminated with DDT.[14]

In most cases the PRD requested that manufacturers voluntarily take offending products off the market. According to PRD assistant director for enforcement Lowell Miller, corporations usually cooperated. Miller had worked for the Rural Electrification Administration and the Federal Trade Commission before joining the ARS in 1967. Before the Fountain subcommittee, he offered the example of a label on a shipment of 25 percent parathion that "bore no warning or caution labeling whatsoever." The ARS requested that the manufacturer recall the shipment and check for other labeling violations. Although the mislabeled parathion was recalled, the fact that such a toxic chemical exited the factory without a proper warning demonstrated a grave breakdown in the labeling process. Until January 1968 no guidelines existed for prosecutions of such violations, and in 1969 the ARS had only one employee assigned this task. Although the ARS was required by statute to publish notices of judgment when products were seized, none appeared from 1965 to 1969.[15]

Unclear or inadequate labels put users in jeopardy, as was sadly illustrated by a case involving Manuel Velez-Velez and Jaime Ramos-Sanches. In 1957, the Hubbard-Hall Chemical Company registered a 1.5 percent parathion dust with a label that did not include a skull and crossbones. The approved label did warn users to wear protective clothing, gloves, and a mask and to wash thoroughly before eating or smoking. In 1959, Velez-Velez and Ramos-Sanches, who came from Puerto Rico, worked on a Taunton, Massachusetts, farm, where they applied parathion four times in mid-August. Neither man wore the protective clothing that was available. Late on August 14 both men took sick, became semicomatose, and, after being hospitalized, died. Physicians targeted parathion poisoning as the cause of death, and a jury found that the Hubbard-Hall Chemical Company label did not adequately warn Velez-Velez and Ramos-Sanches of parathion's toxicity. No question emerged as to whether the PRD had approved the parathion label without a skull and crossbones.[16]

Crop duster pilots and crews relied upon labels, for they knew that their lives depended upon their undiminished skill. Even though most pilots handled deadly pesticides with respect, circumstances conspired to cause accidents. On August 29, 1961, Jose G. Gonzalez, a twenty-eight-year-old pilot, made three flights in Lee County, South Carolina, applying Folex, a plant defoliant. A Mexican citizen attending medical school in Santa Monica, California, Gonzalez was in his second summer dusting crops to pay his way through school. Since his loader was ill, Gonzalez loaded six bags of Folex (some three hundred pounds) into the hopper as the propeller blew the dust behind him. The label lacked a poison notice, an antidote, and a skull and crossbones indicating deadly toxicity. Gonzalez had earlier drawn a map of the farm layout and noted obstructions. As he made his third flight around 7:00 that morning, Gonzalez circled the field checking for obstructions and then started his first "swath run," the actual spraying. Suddenly he became nauseated and short of breath; he also had stomach pain, saw black spots, and felt incapable of making decisions. He attempted to pull his Piper J3 out of the run but hit a wire, stalled, and crashed. The plane hit the ground nose first and flipped over, whereupon the hopper ruptured and spilled Folex over him. He was left upside down, hanging by his seatbelt. Gonzalez suffered a broken left leg, a broken ankle, an injured hand, two broken teeth, and numerous lacerations. His vision was poor, and he was vomiting. Nearby field workers called an ambulance, which rushed him to Lee County Memorial Hospital in Bishopville.[17]

At first Dr. H. E. Farver treated Gonzalez's visible injuries, scheduling an operation that afternoon to insert a rod into his broken leg. But Gonzalez's condition suddenly worsened. "This patient's condition was one of rapid deterioration," Dr. Farver recalled. "His blood pressure fell to nothing, his pulse became barely palpable, he started vomiting copiously and continued to vomit a great deal." He also suffered excruciating pain. Dr. Farver gave Gonzalez intravenous fluids and blood transfusions; the doctor also tried other measures, "none of which seemed to alleviate or improve the shock that [Gonzales] was exhibiting." Dr. Farver inquired about the "powdery substance" that covered Gonzalez and sent for the Folex label, but it contained no antidote. Neither local physicians nor the Columbia Poison Center had information on Folex, so he called the manufacturer, the Virginia-Carolina Chemical Corporation in Richmond. The company's toxicologist was not available, but a company official suggested he call the Duke University Poison Center. Between caring for Gonzalez and making calls, it was after ten o'clock in the evening—twelve hours after Gonzalez arrived at the hospital—before a chemist at Clemson University suggested atropine, which was the antidote for organophosphate poisoning. Upon being given atropine, Gonzalez improved but then had a relapse. After several more days of treatment, he came out of shock. "I was in awful pain, my stomach, I thought there was a ball of fire in it, very unpleasant sensation," he recalled. He also developed pustular blebs—blisters filled with pus—on his leg, hip, and abdomen where he had been prepared for surgery. The doctors drew blood for a cholinesterase test and sent it to Dr. Wayland J. Hayes Jr. for analysis. Hayes replied "that the test showed that it fell in the lower limits of normal" and "that the finding can not be taken as proof of the poison. It is consistent, however, with the presence of mild poisoning." Gonzalez's leg was placed in a skeletal traction system, but when this treatment did not succeed, Dr. Farver set his leg with a metal pin on October 20. Gonzalez remained in the hospital from August 29 until November 17.[18]

FEDERAL AVIATION AGENCY

UNITED STATES OF AMERICA

THIS CERTIFIES THAT

JOSE GOMEZ GONZALEZ

SOUTH HARRINGTON AVENUE

LOS ANGELES 64, CALIFORNIA

7-16-36 | 150 | BLACK | BROWN | MEXICO

HAS BEEN FOUND TO BE PROPERLY QUALIFIED TO EXERCISE THE PRIVILEGES OF

COMMERCIAL PILOT

RATINGS AND LIMITATIONS

AIRPLANE SINGLE ENGINE LAND

Jose G. Gonzalez

SIGNATURE OF HOLDER

BY DIRECTION OF THE ADMINISTRATOR

DATE OF ISSUE 7-8-60

Jose Gonzalez's pilot's license.

Gonzalez *v.* Virginia-Carolina Chemical Company *trial transcript, container 169, AC874, Exhibit P-1, National Archives and Records Center, Southeast Region, East Point, Georgia.*

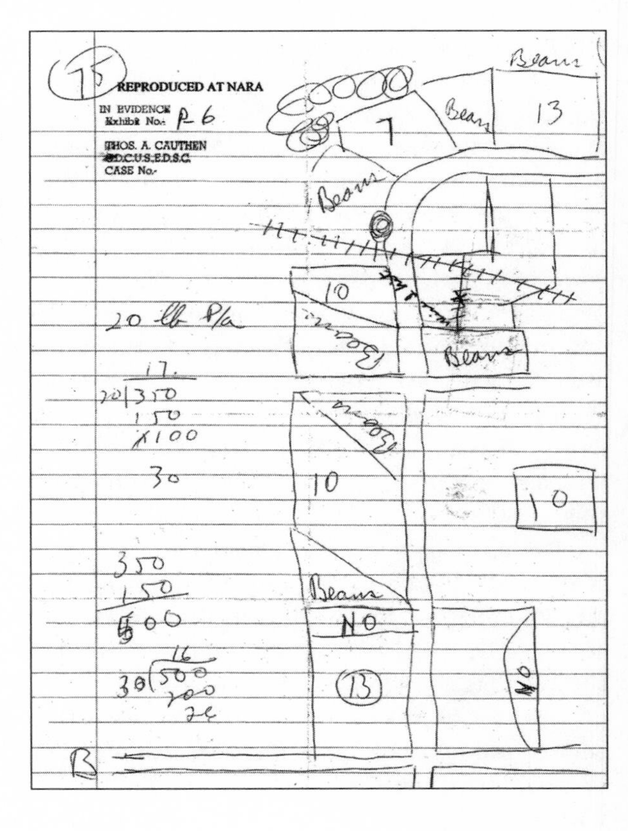

Jose Gonzalez's map.

Gonzalez *v.* Virginia-Carolina Chemical Company *trial transcript, container 169, AC874, Exhibit P-6, National Archives and Records Center, Southeast Region, East Point, Georgia.*

After several follow-up visits to Bishopville doctors, Gonzalez went home to Mexico and began physical therapy. He returned to Los Angeles in June 1962, where he began a job as a short order cook and continued his medical education that fall. In May 1963 he received a general examination from Dr. R. M. Gonzalez in Gardena, California. "I am extremely nervous all the time since the accident," he reported. His left knee, ankle, and hip ached, his right hand was weak, and he had a lump on his right wrist. His tardily set left leg was an inch and a half shorter than his right.[19]

Jose Gonzalez sued the Virginia-Carolina Chemical Company for damages sustained because the Folex label did not properly warn of dangers or contain an antidote. Judge C. C. Wyche heard the case without a jury. The attorney for the chemical company began with a blanket denial that Folex was dangerous. "We propose to show that the toxicity of this material is about in the range of aspirin, perhaps a little more so," he argued. "Everything is toxic, even water, under certain conditions," he shamelessly claimed. Prosecution witness William Wannamaker III disagreed. The active ingredient in Folex, the chemist testified, was the highly toxic tributyl phosphorotrithioite, which could be absorbed through the skin, by inhalation, and through the eyes. Wannamaker had attended the Chemical Warfare School at Fort McClellan, Alabama. "We studied how to kill people in broad terms and this was done mainly with organic phosphorus compounds," he explained. Judge Wyche observed that the effects of Folex were cumulative, and he concluded that Gonzalez crashed because of a "typical case of cholinesterase inhibition," which he defined as "the slowing of nerve impulses and lack of muscular coordination." The judge had discovered that the original label submitted to the PRD had included the words "avoid prolonged or repeated inhalation of dust," but that the company had deleted "prolonged and repeated" from the label. Nor had company chemists done research on its toxicity. The Virginia-Carolina Chemical Company, Wyche ruled, "was negligent in distributing a dangerous poison without adequate tests and without adequate warning of its toxic effects and in failing to publish and/or have available a protective antidote for the toxic substance, and such acts of negligence contributed as a proximate cause to plaintiff's injuries and damages." Judge Wyche allowed Gonzalez to collect $40,000 in damages.[20] The PRD had apparently approved the Folex label without making the company submit vital information; the label listed neither an antidote nor a warning of toxicity. It was impossible to determine what the Virginia-Carolina Chemical

Company had submitted. There is no indication that the ARS took action against the Virginia-Carolina Chemical Company, either to make it recall Folex or revise its label.

A South Carolina poisoning case also began with a problem label. On May 18, 1964, twenty-eight-year-old Edward V. Griffin Jr. was taking inventory in a pesticide warehouse. He had graduated from North Carolina State University in 1959, married Anne Gibson Page, and several years later moved to her hometown, Lake View, South Carolina, to manage a farm supply business. That Monday afternoon he made three trips to the pesticide warehouse. Later he talked with two workers in the fertilizer warehouse. His shirt was wet, and he explained that a bag of 1 percent parathion dust had broken open and that he had washed the dust off with soap and water. One of the men suggested that Griffin might want to call a doctor. Griffin went home at 6:00 P.M. and watered the flowers. Anne Griffin, who was pregnant with their third child, called him to supper. While eating, he became ill. Anne called for an ambulance, which rushed him to the hospital at Mullins, twelve miles away. He arrived unconscious, and Dr. Daniel Pace quickly but unsuccessfully administrated atropine. Griffin died about an hour later. The next morning workers found a burst bag of 1 percent parathion dust in the warehouse.[21]

Anne Griffin sued both for pain and suffering and for wrongful death. The cases were consolidated, and Judge Robert W. Hemphill heard them without jury on June 11 and 12, 1969. Anne Griffin's lawyers insisted that Edward Griffin, who had attended pesticide conferences at North Carolina State College and understood pesticide toxicity, had followed the instructions on the label and had washed himself thoroughly. Despite parathion's toxicity, the label—approved in 1949 by USDA and never revised—contained no skull and crossbones, nor did it mention an antidote. Planters Chemical Company enlisted Mitchell Zavon as an expert witness. Zavon admitted that "in all probability" Griffin died from parathion poisoning. Judge Hemphill declared the label "incomplete and inadequate" and faulted Planters Chemical Company, ruling that its responsibility for an accurate label went beyond ARS approval. Anne Griffin won a settlement for pecuniary loss, loss of companionship, and mental shock and suffering.[22] But it was not the clarity or the color of the label that doomed Edward Griffin; it was misinformation. From February 1955 through February 1968, the ARS had not reported a single violation of registration rules to the Justice Department for prosecution. There is no record that it called Planters Chemical Company into account.

Edward V. Griffin Jr.

Anne Page Griffin *v.* Planters Chemical Corporation, *civil case 69-170, accession number 72E1507, box 169, location G0781354, National Archives and Records Center, Southeast Region, East Point, Georgia.*

Burst parathion bag.

Griffin *v.* Planters Chemical Corporation, *civil case 68-170, accession number 72E 1507, box 169, location G0781354, National Archives and Records Center, Southeast Region, East Point, Georgia.*

Handling parathion bags, as the Griffin case demonstrated, presented grave dangers. In the mid-1970s, a worker at an experiment farm in Georgia moved ten bags of 10 percent parathion. Later he lost consciousness and "was on the floor, shaking, glassy-eyed, and foaming at the mouth." He never recovered and died a short time later. Nearby was an open parathion bag. It took very little parathion to do tremendous damage, yet clear labels and packages strong enough to withstand handling did not become a regulatory issue.[23]

The corporations that manufactured, formulated, and distributed pesticides were also prone to errors. Farmers often complained that insecticides did not perform as advertised. In September 1962, some cattle at Lubbock, Texas, died after being sprayed with Chemagro Corporation's Co-Ral, an organophosphate insecticide. R. D. Radeleff, the veterinarian in charge of the animal disease and parasite research center in Kerrville, Texas, conducted a thorough investigation. The cattle showed symptoms of acute poisoning, recovered, and then had relapses. Three hundred died out of the eight thousand affected. The poisoned cattle would not eat, became gaunt, and suffered from diarrhea and abdominal pain. The most badly affected "move slowly, somewhat stiffly, dragging the hind toes, sometimes knuckling over." Radeleff cooperated with Chemagro scientists and discovered that due to a formulation error, Co-Ral samples measured a twofold increase in toxicity. Chemgro recalled one hundred lots of Co-Ral from distributors. Radeleff worried that public pressure was increasing to "suspend the labeling and shipping permits (registration) and recommendation of Co-Ral." The problem, he stressed, "is quality control failure rather than a change in the basic chemical in Co-Ral. Properly manufactured, Co-Ral can still be a useful compound."[24]

Radeleff, like some other scientists in the ARS, attempted to learn from such incidents. In March 1963 he fretted that the ARS would "progressively drop behind the times in studying the reproductive and toxic effects of these materials unless our number of researchers is markedly increased." He considered "this new series of compounds" capable of "inflicting more insidious damage than has thus far been attributed to pesticides." Radeleff concluded that "we have been exceedingly lucky in our system, dictated by limitation of personnel, and funds." He had recently watched a "CBS Reports" program that featured Rachel Carson debating government and industry representatives. The program, he concluded, "certainly pointedly emphasized the relatively small knowledge we have in relation to what should be known."[25]

The small ARS inspection teams were hard pressed to scrutinize companies and ensure that pesticides contained the material listed on the label, that labels contained required information, and that products met standards. There is no way to determine the number of inaccurate or insufficient labels that cleared the PRD and ended up on bags of toxic chemicals; nor is it known how many chemical companies altered labels. But innocent victims paid a price for such labeling errors. Pilots, loaders, and workers moved through a chemically tinged environment, and their experience and common sense, more than labels or expert guidance, allowed most of them to survive. As with radiation damage, the long-term chemical effects on humans remains difficult to measure. Whether this exposure will ultimately turn out to be benign or will visit effects on later generations is a question still blowing on the wind.

Farmers sometimes failed to read labels, with drastic consequences. On June 28, 1965, Elton Wilson, a diary farmer who lived near Rives, Tennessee, sprayed his herd with Purina Dairy Spray Concentrate to control flies. His hired hand mixed the chemicals and began spraying. When that batch was used up, Wilson "reached up into the loft above the feed room and got a bottle of spray," mixed it, and continued with the job. In a few minutes the hired hand "came running into the barn and reported that he saw the herd bull stagger and fall and when he went over to see about him he found him dead." Wilson saw other cattle "staggering and falling down." When the veterinarian arrived, he noticed an empty bottle of highly toxic Purina Diazinon Spray, which Wilson had mistakenly used to mix the second batch of chemicals. The veterinarian administered the antidote, but thirty cows plus the bull died. "Mr. Wilson stated that he would never buy any more Diazinon Spray," a USDA inspector reported, "and admitted that it was a mistake on his part in using the spray and not reading the labeling on the bottle." Meanwhile, Wilson was unable to sell the milk from his remaining cows until reports on diazinon residue were completed.[26]

The PRD sometimes approved labels with contradictory information and often simply used label copy from pesticide formulators. At Congressman L. H. Fountain's 1969 hearing, committee counsel James R. Naughton read from such a label: "Use in well-ventilated rooms or areas only. . . . Do not stay in room that has been heavily treated. Avoid inhalation." On the other side it read, "Close all doors, windows and transoms. Spray with a fine mist sprayer freely upward in all directions so that the room is filled with vapor. If insects have not dropped to the floor in three minutes, repeat spraying. . . . After ten minutes, doors and windows may be opened."

Naughton pointed out the dangers of such labels and quizzed PRD head Harry Hays about accidental poisonings. Hays was unfamiliar with statistics from the PHS, which reported 4,000 child and 1,000 adult poisonings in 1968. Naughton claimed that such poisonings were underreported and that 50,000 was a more accurate estimate. The hearing brought out the PRD's laxity in policing labels, in recalling products, and in following up with manufacturers.[27]

The PRD's inactivity regarding lindane vaporizers epitomized its close ties with corporations and raised disturbing questions about its concern for human health. In 1948 the English Aerovap Corporation registered a DDT vaporizer; a year later, as flies gained immunity to DDT, it substituted lindane. The vaporizer consisted of a heating coil encased in a cone-shaped holder. Lindane pellets inserted into the heated chamber produced odorless and colorless vapors that killed flying insects, while lindane residue settled on walls and flat surfaces. In 1950, the PHS and the FDA approved lindane vaporizers, but they agreed with the PRD that the label should carry a "Not for Home Use" warning. In 1952 an interdepartmental committee on pest control recommended against permitting vaporizers "in rooms or areas where food is served, processed or stored." The ARS ignored the recommendation. At the same time, USDA meat inspection rules forbade lindane vaporizers in meat processing facilities or slaughterhouses. The ARS also approved "one shot" fumigators, which could be used in homes under strict conditions that protected pets, food, and people. The PHS continued to raise questions about safety, but the PRD ignored them. By 1968 there were roughly one hundred registrations for the two types of lindane vaporizers. Millions had been sold. In 1965 and 1966, the PHS and a committee of experts formally opposed lindane vaporizers. One side of the lindane vaporizer label, it turned out, recommended use in restaurants, bars, and offices, while the other side cautioned against inhalation and food contamination. In March 1969, a three-person advisory panel connected lindane with blood cell dyscrasia. The panel concluded that using lindane "where people are exposed or where food intended for human consumption is stored or served creates a hazard which is potentially detrimental to human health, and it is our opinion that the sale of lindane pellets intended for use in vaporizers should be discontinued."[28]

Reports from across the country indicated that lindane vaporizers were dangerous to some people, although others seemed immune to its effects. The lindane itself was definitely toxic. In August 1957, Mrs. E. H. White's

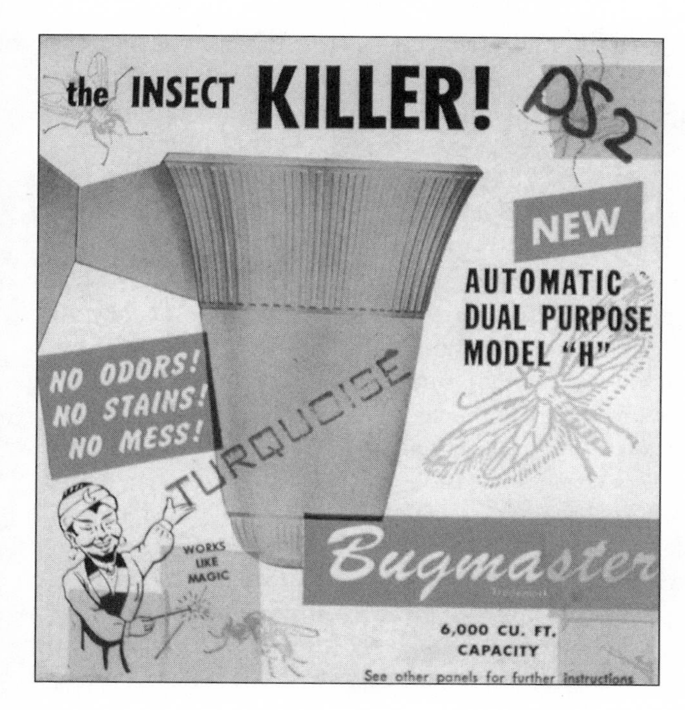

Bugmaster lindane vaporizer box. *Landmark litigation files, 1970–74, EPA, RG 412, box 7, NARA.*

youngest child swallowed some lindane pellets and went into convulsions before she could get him to a doctor. "The pellets were marked poison," she explained, "but do not show any antidote" on the package. The doctor in Monroe, Louisiana, had no information on lindane, nor did the parish library. Mrs. White's child had not fully recovered seventeen months later. PRD chief Justus C. Ward wrote to Mrs. White expressing "regret" at the child's poisoning but insensitively chiding her for leaving the lindane pellets within his reach. Ward also denigrated the physician for not quickly finding an antidote. His highly technical reply explained away not only the need for a skull and crossbones warning but also any PRD culpability. He thoughtfully listed several poison centers and suggested that Mrs. White's physician might make a note of them. There was no mention of a label review, let alone an admission that lindane had been the subject of several studies that revealed it to be injurious to human health.[29] The tone and content of Ward's reply epitomized that of a self-righteous and callous bureaucrat.

One needed not swallow lindane to suffer its effects. In August 1959, sixty-one-year-old Harold W. Lawson, a Little Rock, Arkansas, druggist, lit several dozen Vaporette lindane candles in his drugstore to kill roaches and waterbugs. When he reentered the store to relight some of the candles, he breathed the lindane vapor. The next morning he was ill with diarrhea and nausea; he then fainted and fractured his hip. After a hip operation, he suffered "acute respiratory obstruction," "renal failure," and "electrolyte imbalance" and was transferred to the University of Arkansas Medical Center. Dr. Ben I. Heller, professor of medicine, observed that "the gamma isomer of benzene hexachloride is extremely toxic to the bone marrow." Heller described Lawson as "poorly nourished," and subsequent research would suggest that malnutrition enhanced the toxicity of lindane in laboratory animals. In March 1960, the Arkansas Board of Health held hearings on the safety of Vaporette candles. One of the expert witnesses, James C. Munch, reported the proceedings to the PRD, suggesting that it examine the hearing transcript and review evidence relevant to lindane labels. Although it was likely that lindane set in motion Harold Lawson's illness, Dr. Heller admitted, "There is no way of proving this conclusively."[30]

As the ARS bureaucracy ignored data on lindane's toxicity and continued to allow inadequate labels, Louis Kamansky and his wife, who lived in Upland, California, desperately attempted to get lindane vaporizers off the market. Their daughter, Debra Joy, had played at a girlfriend's house nearly every afternoon for three years. There was a Bug Death lindane vaporizer in the room where the girls played, and several other vaporizers were scattered around the house. In August 1960 Mrs. Kamansky noticed that seven-year-old Debra was nervous and had developed black and blue marks on her legs, arms, and buttocks. After visiting several other doctors, the family consulted Dr. J. Philip Loge, a hematologist who taught at the UCLA Medical Center. He diagnosed her condition as aplastic anemia and observed that her only history of toxic exposure was to lindane. Her health deteriorated, and in June 1961 she received a blood transfusion and was given steroids. Then eight years old, Debra had gained weight, her hair had coarsened, and she had spots all over her body. She was listless, languidly watching TV and drawing pictures. At one point that summer her condition worsened when the family ate at a restaurant that, unknown to them, had a lindane vaporizer.[31]

Mrs. Kamansky was outraged that lindane vaporizers were used not only in homes, but also in restaurants and other public spaces. She had appealed to California authorities, written to her senators and to the president, appealed to the PHS, pondered press exposure, and considered legal action. When D. W. Dean, an ARS field representative, interviewed the Kamanskys, he lamely claimed that home use of lindane vaporizers was not federally approved and that the products were seized in interstate commerce. Dean assured the family that the ARS would continue to seize vaporizers and "probably" would ban them in restaurants. Such assurances did not placate Mrs. Kamansky.[32]

On June 19, 1961, Dean interviewed Dr. J. Phillip Loge, who gave a review of Debra Kamansky's case. "Because of the background of exposure," Dean reported, "he feels that lindane is responsible for the girl's condition." Loge had found six other cases of anemia conditions "due to gamma isomer of benzene hexachloride," that is, lindane. He had also been interviewed by attorney Melvin Belli about the Kamansky case. Dean judged Loge as "a very competent physician" who "would give sincere consideration to the problems of his clients," but also "one who is well aware of the complications that can arise from medical legal law suits." Loge would not testify that lindane caused Debra Kamansky's illness, Dean learned, "without further investigations and tests of a type that he may be unwilling to undertake." He had examined Debra's playmates and had found no blood problems with them. Dean also learned of four other cases in which lindane vaporizers were suspected of causing illness or death.[33]

Mrs. Kamansky had pointed out to Dean that the cautionary labeling on Bug Death vaporizers, which were put out by the B-D Products Corporation, were not visible to the buyer. The lindane pellets with the warning, she complained, were glued to the inside bottom of the container. Nearly everyone Dean interviewed had reservations about lindane vaporizers. Dean found "evidence that certain susceptible people develop aplastic anemia after considerable exposure to lindane," and he recommended that the PRD reevaluate its registration policy for the chemical. He strongly recommended that the PRD seize lindane vaporizers that were targeted for home use or involved in interstate commerce. In a letter to Dr. Loge in August 1961, Wayland J. Hayes Jr., who had seen Debra Kamansky's medical reports, admitted that lindane or benzene hexachloride had been associated with blood dyscrasia more than any other chemical, but he argued that establishing a direct cause-and-effect relationship would require further research.[34]

Mrs. Kamansky's activity in behalf of her daughter put the ARS under pressure to make a response. "I have the deepest feeling of urgency in this matter," Mrs. Kamansky wrote to California senator Thomas H. Kuchel in

May 1961, "knowing the tragic effects of the merchandising of this poison on my own daughter, who appears to be failing." When the ARS responded to Kuchel's inquiry, it made no mention of potential lindane-related health problems, which had first been raised in the early 1950s. "Lindane was studied extensively by pharmacologists of the Food and Drug Administration and other scientists," E. D. Burgess wrote in June 1961, "and was found to be safe in foods at the levels shown in the enclosed photocopy of its list of tolerances." Burgess denied that lindane vaporizers had been registered for continuous home use and assured Mrs. Kamansky that the ARS was conducting further studies. ARS officials bent Dean's interviews and reports to minimize lindane dangers. "The medical authorities contacted could not be sure that exposure to lindane was the actual cause of Debra's illness," USDA assistant secretary Frank J. Welch reported, "although they felt that it might have contributed to it." Welch reiterated Burgess's claim that the PRD had not registered vaporizers for home use, but then he added, "except where the registration was granted under protest." There were no statistics on how many products entered the market through that door. Meanwhile, Debra Kamansky continued to fail. "As a result of treatment for severe aplastic anemia," a report stated in November 1961, "she is now pitifully disfigured with swelling on checks, abdomen, and back and face, arms and body covered with petechial spots." After fourteen months of failing health, Debra Joy Kamansky died on November 28, 1961.[35]

The Kamansky case raised serious questions not only about the veracity of ARS statements but also about its concern with legitimate questions of chemical toxicity—questions that had hounded lindane vaporizer use. On April 29, 1969, just a few days before Congressman L. H. Fountain's hearings, the PRD announced that it was beginning action to cancel lindane registrations. Despite a 1953 FDA report on toxic residues and numerous reports of lindane-related illness and death, the PRD had not acted earlier, it claimed, because it did not have scientific data to establish a hazard. In 1969, the PRD at last conducted a five-day test, establishing that lindane vaporizers left residues on exposed food. For sixteen years, then, the PRD failed to conduct tests or evaluate scientific evidence on vaporizer dangers. Claiming that lindane vaporizers were not registered for home use while also admitting that they were registered under protest was disingenuous. Instead of taking action to remove an obviously dangerous product from the market, the PRD simply continued to allow lindane registration. The mounting toll of lindane-related illnesses showed how thoroughly the ARS ignored its responsibility for protecting human health. Between 1955 and 1970, there were sixty reported cases of lindane poisoning and at least eight deaths. Despite the trail of sickness and death, Mitchell Zavon never wavered in his support of lindane vaporizers. Testifying before an EPA panel on June 16, 1971, he stated that no warning label was needed to protect old or young people. "Nor do I know of any evidence that would indicate that infants, children or elderly persons are particularly at risk from lindane or Lindane vaporizers," he stated.[36]

Accurate accidental poisoning statistics were difficult to obtain, for reporting was hit or miss. Until 1966, the ARS maintained no reporting network for poisonings and relied upon scattered reports from other government agencies, as well as state and local sources. Complaints, inquiries, and reports in ARS files represented only a fraction of national cases. In 1968, for example, the ARS received an estimated 150 to 175 accidental poisoning reports; the PHS received some 5,000, "of which approximately 4,000 involve children under 5." At Congressman Fountain's 1969 hearings, counsel James S. Naughton discovered that the PHS estimated that the total number was actually eight to ten times greater than it received. When PRD chief Harry Hays could not estimate how many of the 5,000 cases each year resulted in death, New York congressman Benjamin S. Rosenthal was incredulous. "How can you operate without knowing these figures? How can you even continue to certify pesticides without knowing these figures?" he asked, sternly adding that "it is insulting to this committee for you to appear before us and not have that information."[37] As this exchange suggests, the two 1968 GAO reports, the Fountain hearings in May and June 1969, and a subsequent House report stripped away the camouflage and revealed the intricate association between chemical companies and the ARS.

After the GAO reports, the ARS moved desultorily. It issued a new inspection manual, required manufacturers to submit samples for testing, and asked for product shelf-life studies. The ARS continued to insist on "recall actions, rather than seizure actions." The USDA general counsel reviewed the dispute between the ARS and the PHS over lindane vaporizers but observed that no agreement had been reached—in large part because no meeting had been held to resolve the matter. Finally, the USDA inspector general's report insisted that PRD administrator Harry Hays reorganize and strengthen the division.[38]

In April 1969, George Irving Jr. reported to Secretary of Agriculture Clifford M. Hardin that the ARS had taken several positive steps since the November 1968 GAO report. Medical experts had met to discuss residue studies, the ARS had promised to set up meetings with the Department of Interior and the HEW, and the PRD had conducted a study of lindane residues. The study, Irving informed the secretary, indicated "adulteration of food at the third, fourth, and fifth day of exposure." He asserted that it "will provide a basis for further action on the registration policy for lindane pellets."[39] Even at this point, however, Irving did not indicate what kind of action he might take. Meanwhile, lindane vaporizers remained on the market.

While the PRD's inaction on lindane vaporizers revealed its disregard for human health, its treatment of the vapona pest strip displayed stalling, favoritism, and outright incompetence. The vapona story began in 1955, when PHS research scientists in Savannah discovered the insecticide DDVP, also called vapona, and placed it in the public domain. Shell Chemical Company seized the patent rights. In January 1963 Shell applied to the PRD for registration of a DDVP/vapona pest strip that killed flies by releasing toxic vapor. A month later, a PRD pharmacologist advised Shell that the DDVP label should contain a poison warning and a skull and crossbones. On March 6, Shell representatives visited the PRD, and the next day John S. Leary, the PRD's chief staff officer for pharmacology, overruled the pharmacologist's objection and registered Shell's DDVP pest strip without the suggested warning. For the next two years, Leary attended meetings on DDVP pest strip safety with representatives of Shell and other government agencies, including HEW personnel (who opposed not only DDVP but also lindane and pyrethrum vaporizers). Still, the PRD claimed it did not have enough scientific evidence to cancel DDVP registrations. In November 1965, the PRD approved eight additional DDVP registrations to Shell or its affiliates. The PRD refused to participate in a 1966 PHS study that opposed vapona registrations. In October 1966, PRD director Harry Hays met with three medical consultants and handed them Leary's memorandum favoring vapona registration rather than the PHS report opposing it. Leary's memorandum also suggested that vaporizers be labeled air additives rather than air polluters. On November 14, 1966, Leary resigned from the PRD effective December 31 to take a position with Shell. On December 6, not waiting for the door to revolve, he wrote a memo to Hays arguing that the PHS report "serves no useful purpose" and that it "should be set aside as inadequate justification for any change in the registration status of these methods of pest control." After assuming his position at Shell, Leary attended a PRD meeting in July 1968 in order to contest a pest strip label change that would warn of dangers to infants and ill and aging people. Leary submitted a document, "Vapona Human Study Evaluation: Literature Review," which concluded that DDVP "offers no hazard to human health."[40]

As the DDVP debate developed, Secretary of Agriculture Orville Freeman in July 1965 appointed a seven-member task force to examine PRD effectiveness in registering pesticides. Harry Hayes, then at the National Academy of Sciences, became its chair. The rest of the panel included four USDA members and a California agricultural official, joined by Dr. T. Roy Hansberry, a Shell research director. When Hansberry submitted a conflict of interest form, the USDA personnel office falsely declared that the ARS had no "official business" with Hansberry's firm. The form never reached the office of the general counsel. Shell, of course, had numerous registrations with the PRD—over 250, according to 1969 tabulations. When the task force met on July 8 and 9, 1965, executive secretary Harold G. Alford took notes and reminded the committee that corporate formulas could not be discussed in Hansberry's presence. When Alford arrived for the next meeting, Hays dismissed him and no official notes were taken. Alford's notes from the first meeting disclosed that the committee discussed Shell registration cases. Two of Shell's applications for DDVP strips were pending at the time.[41] The conflict of interest process completely broke down, allowing a Shell employee to infiltrate the registration task force.

After testifying in the *Lawler* case as an expert witness, Mitchell Zavon increased his consultantships, defending pesticides at every opportunity. Identifying himself as "a physician," a professor of industrial medicine at the Kettering Laboratory, and a director of occupational health with the Cincinnati Health Department, he testified about color additives before the House Committee on Interstate and Foreign Commerce on April 5, 1960, a month after testifying in the *Lawler* case. From March 1963 until June 1969, Zavon was a PRD consultant; at the same time, he served Shell Chemical Company as a medical consultant. His years with the PHS, 1949 to 1957, alerted him to its concern—or lack thereof—regarding pesticides. Conflict of interest laws did not prohibit Zavon, as a PRD consultant, from conducting a firm's business with the government, but the laws did prohibit him from participating in matters in which the firm had a financial interest. Zavon's association with Shell raised a flag, and the USDA office of the general counsel suggested that the pertinent law be brought to Zavon's attention. Even while consulting with the PRD, the Fountain report noted, Zavon "individually or with others, conducted a number of scientific studies used by Shell to support its contention that DDVP was safe." Scientists at the Kettering Laboratory, where Zavon was employed, also did DDVP studies. After consulting with John Leary in 1964, Zavon headed a study of food exposed to DDVP strips in an Ohio restaurant; he sent the data to a Shell laboratory for analysis. Unlike other laboratories that did similar studies, Shell found no DDVP residues.[42]

When not advising the PRD or conducting research, Zavon was attending meetings relating to DDVP strips. As a Shell consultant in April 1965, Zavon met with John Leary, two PHS staffers, and three other Shell representatives. In June 1967, as a PRD consultant, Zavon met with the PRD advisory committee. Both meetings discussed DDVP strips. Meanwhile, Shell attempted to neutralize the PHS study that revealed vaporizer dangers. Zavon denounced the PHS study and championed his own DDVP research. Using his position as assistant health commissioner in Cincinnati—but not his Shell association—Zavon called for another PHS study. He continued to lobby against the vapona label change that would warn of health risks to infants, invalids, and aged people. In addition to making phone calls, he visited PHS headquarters in Atlanta in August 1967. In September, he negotiated a compromise label: "Do not use in rooms continuously occupied by infants and infirm individuals." The PHS accepted this wording, but Harold G. Alford at the PRD insisted on omitting the word "continuously." On December 11, Shell's John Leary drafted a letter insisting that new evidence proved that the label required no health warning. "It was not always clear," the *Washington Post* reported in 1969, "whether Zavon was representing the government or Shell Chemical."[43]

In December 1968 the GAO issued its report on vaporizers. It dismissed Shell's additional data and insisted that the label must be changed. In January 1969, even as it protested that it did not agree with a cautionary statement, Shell at last agreed to include a warning on pest strip labels. As late as the summer of 1969, however, counsel James Naughton asserted that "you can look in many stores in the Washington area and find large stocks of Shell No-Pest Strip which don't bear that warning." When Naughton asked Harry Hays why these products were not seized, Hays replied that "it didn't appear to me to be practical in this instance." Naughton asked, "What was impractical about it? Shell is a big company. They could find somebody with a rubber stamp to go around and put the warning on the packages." Finally, the USDA referred the conflict of interest cases of

Hansberry, Leary, and Zavon to the Department of Justice. In August 1969, Justice found no conflict of interest in Hansberry's case. Presumably Leary and Zavon were also cleared.[44]

Even as Shell danced with the ARS over label changes, Francis Silver, a chemical engineer from Martinsburg, West Virginia, rebuked the ARS's George L. Mehren for claiming that customers were protected if they followed label directions. "Most of the injury is not from violating labeling instructions," Silver claimed in June 1967, "but from following labeling instructions of improperly designed products." He mentioned numerous home and janitorial accidents. "Perhaps the most glaring example of this is the current heavy promotion of the DDVP strip for insect control in the home and other inhabited buildings," he fumed. "The home owner is being coaxed to try to make the air in their homes too toxic for insects to live in them." Even registering DDVP strips, he charged, "makes a mockery of your claim of protection. The fact that this application is permitted to be promoted over TV in the Washington area, right under the nose of DA, FTC, PHS, and Congress clearly reveals that consumer protective law does not protect." Rather, he concluded, it was "nothing more than a public relations device for dispersing the public's natural caution and promoting sales."[45]

The GAO reports and the Fountain hearings prompted the ARS to begin action on Shell's no-pest strip. On September 9, 1969, ARS staffer Harold Alford warned Shell that research revealed that DDVP strips produced residues that adulterated food. Labels needed to carry a warning: "Do not use in kitchens, restaurants or areas where food is prepared or served." Alford's survey of the Washington area, like Naughton's, revealed stocks of DDVP that did not have the warning about health risks to infants, invalids, and aged people. All stock should be relabeled, he cautioned. Otherwise steps would be taken to cancel registration. On September 24, Shell's W. G. Appleby countered that the company was "developing additional information" that would clear vapona for use in food areas and would rule out a need for the revised label. Appleby threatened that he wanted to resolve the issue without resorting to an advisory committee or public hearing.[46]

While the PRD dallied on lindane vaporizers and vapona pest strip warnings, it acted with alarming haste when a Shell rival marketed a product similar to its pest strip. Aeroseal, a small Pennsylvania company, had formerly marketed a strip using Shell's vapona. When it redesigned the product and no longer relied upon Shell as its supplier, Shell took samples to the PRD, which immediately tested Aeroseal's product, declared it faulty, and recalled it. It took a mere four days in March 1969 for the PRD to complete the process. It insisted that the Aeroseal product "bore no resemblance to the product that Aeroseal had previously registered with the Division." Members of the Fountain committee asked why the Aeroseal product, with little testing and no consumer complaints, could be taken off the market in four days while the PRD had dawdled for years with Shell's refusal to change its warning label to protect infants and the elderly. James Naughton observed that the two products "had exactly the same active ingredients and practically the same, if not identical, directions for use," so he wondered "what made one an imminent hazard when the other didn't give you enough concern that you even thought about having the warning put on the boxes in the stores?" PRD staffers argued that the Aeroseal product was different, but none of them could recall another case that was completed so rapidly. Two months after the May 7 hearing, Aeroseal again attempted to register its product, arguing that it released less DDVP than the Shell strip. Since Aeroseal was manufactured in Congressman George Goodling's district, ARS staffers worried that a negative decision might rile the congressman and gain the attention of the House oversight committee. Goodling pointed out that forty jobs were at stake. The PRD stuck to its decision to demand more tests. In the fall of 1969, PRD also stalled on registering a Busters, Inc., pest strip, citing a series of tests that it deemed to be insufficient.[47]

It was November 13, 1969, before Congressman Fountain sent the USDA the unanimous committee report on ARS deficiencies. He urged Secretary of Agriculture Clifford M. Hardin to give the report his "close personal attention" and asked to be informed about any proposed actions to correct ARS shortcomings. Hardin sent a brief acknowledgment two weeks later. On December 23, Fountain forwarded Hardin a clipping from the *Des Moines Register,* which quoted the apparently puzzled secretary as saying that the Fountain committee report "has not been brought to my attention." Hardin also denied that the ARS had failed to enforce the law. Fountain demanded an explanation. Hardin lamely admitted that he had not read the report at the time of the Des Moines press conference but insisted that he had discussed the issue with USDA staff. "In fact," he explained to Fountain, "we have been actively implementing the recommendations during the past several months." His defensive remark about the ARS enforcing the law, he explained, "was based on the current efforts to bring about the corrections needed and was not intended as a contradiction of the Committee's findings."[48] Obviously, the seriousness of the ARS conflict with Fountain's committee had not been a high priority at the USDA.

The November 13, 1969, House report summing up the Fountain hearings condemned the PRD. In bold print, it concluded: "Until mid-1967, the USDA Pesticides Regulation Division failed almost completely to carry out its responsibility to enforce provisions of the Federal Insecticide, Fungicide, and Rodenticide Act intended to protect the public from hazardous and ineffective pesticide products being marketed in violation of the act." The PRD had not brought a single criminal prosecution despite evidence of continual violations. It had approved products over the objection of HEW, including products that were certain to adulterate food. It had failed to demand clear labels, had made decisions based on inadequate data, had refused to cancel products that might be dangerous, had neglected to remove hazardous products from the market after a cancellation, had failed to establish a way to warn consumers of hazardous products, and had allowed conflict of interest problems with consultants. It was August 1970 before DDVP manufacturers were told to place the warning "Do not use in restaurants, kitchens, or other places where food is prepared or served" or lose their registration.[49] The House report was a portrait of, at best, incompetence, but it also painted a disturbing picture of bureaucratic dereliction. PRD bureaucrats were firmly in the grasp of the chemical industry. There was little left for the PRD not to do.

In the aftermath of the congressional hearings and as pressure mounted to ban chlorinated hydrocarbons, the press reported in June 1970 that some USDA workers thought change had come to the department. "The general feeling around here used to be: Don't offend the pesticide industry," an unidentified USDA official admitted a quarter century after synthetic pesticides arrived on the market. "And when it came to human health, the department was willing to risk a degree of hazard. But now we're taking the position that there must be a big margin of safety for the public." Given the ARS's history, this statement must be read carefully. The most significant aspect of the quotation was not that the department had changed but the confession of how closely it had been linked to the chemical industry. Hansberry, Leary, and Zavon were only three corporate representatives tied closely to Shell. Yet the ARS dealt with dozens of firms, their representatives, and their products. It subscribed to a bureaucratic creed that supinely executed corporate interests, and its actions compromised the nation's health, gave a death sentence to some poisoning victims, and allowed fish and wildlife devastation.[50]

The records of the PRD are not among the ARS material at the National Archives, nor could they be located at the EPA or USDA. Even if we could unearth the missing PRD files and breach Shell's inner sanctum, we might be no closer to truth than the character of Tyrone Slopthrop in Thomas Pynchon's *Gravity's Rainbow.* In one of his paranoid fantasies, Slopthrop imagines an assault on the control room of Shell Mex House in London, searching for Shell's links to the V-2 rocket and to plastics. Instead of a corporate director or indications of flight, Slopthrop finds "only a rather dull room, business machines arrayed around the walls calmly blinking, files of cards pierced frail as sugar faces." He suspects that any evidence he might uncover would be misleading "because the organization charts have all been set up by Them, the titles and names filled in by Them, because Proverbs for Paranoids, 3: If they can get you asking the wrong questions, they don't have to worry about answers."[51] It is fair to ask, however, whether PRD dereliction of duty, consultants' conflicts of interests, or Shell Chemical's suborning of the label process created any feelings of guilt, irresponsibility, or remorse. We do know there are none in the public record.

8
NEO–FIRE ANT ERADICATION

In this connection, I want to point out that in the case of the fire ant eradication program we are laying our collective necks on the block for something which, as I have already indicated, most objective scientists agree is of limited economic consequence.

Louis N. Wise, May 28, 1970

Even as the movement to ban chlorinated hydrocarbons gained momentum and Congressman L. H. Fountain's hearings progressed in the late 1960s, a fire ant eradication revival swept through the South. The fact that the Agricultural Research Service (ARS) even contemplated a second assault on an insect that had co-opted the first by expanding its range and toughening its resistance demonstrated a startling insensitivity to science and history and a careless love for chemical control programs. Despite the marginal economic impact of fire ants, the prospect of controlling millions of dollars and managing another intrusive aerial spraying campaign seduced ARS bureaucrats. Having learned nothing and forgotten nothing, the ARS blundered into the second campaign as it had the first, with inadequate research on both mirex (its chlorinated hydrocarbon of choice) and the fire ant's diet and life cycle.

In 1967, ten years after the first fire ant eradication effort began, a Mississippi state representative sent Secretary of Agriculture Orville Freeman a report from a joint session of the Mississippi legislature entitled, "The Dreaded Fire Ant." It repeated discredited claims from the 1950s: fire ants injured humans, consumed calves, and harmed livestock, among other inventions. Paradoxically, these alleged fire ant depredations more closely fit the effects of agricultural chemicals—such as heptachlor, dieldrin, and aldrin—which did threaten humans, domestic animals, and wildlife. Secretary Freeman, who earlier had condemned the first fire ant crusade as "a classic case of how not to do things," changed his mind after meeting with state commissioners of agriculture in May 1967. "When a State like Georgia," he marveled with veiled condescension and faulty logic, "will appropriate a million dollars to try and do something about it, it must be serious." Freeman conceded that fire ant eradication "may be possible" over ten to twelve years, but he also warned that the bill would come to "as much as $150 million." Before a program could begin, Freeman suggested, more research was needed.[1] With Freeman softened up, southern advocates began to rouse support both at home and in Congress. The ARS ignored Freeman's insistence on research and instead actively lobbied for the control program.

In the summer of 1967, ARS staffers solicited support for fire ant eradication from land grant university scientists and the Southern Plant Board. Most scientists in the land grant network supported the project, but W. A. Ruffin, supervisor of the Alabama division of plant industry, complained that Dr. Dale Newsom "obviously still has a closed mind on the subject" and thought that fire ant eradication was impossible. Ruffin suggested that the Alabama effort needed support from the Farm Bureau, the Cattleman's Association, nurserymen, and other stakeholders. Agribusiness lobbyists quickly convinced Congress to appropriate $5.3 million for fiscal year 1967–68 to spray 2.4 million acres near Savannah, 831,000 acres near Macon and Waycross, and 3.5 million acres in Florida. The ARS also targeted millions of acres in Mississippi, Arkansas, and Louisiana for future treatment.[2]

Perversely, the ARS exhumed its scare tactics from the earlier campaign. Auburn University entomologists complained bitterly about an article in a February 1968 USDA bulletin, which asserted that fire ants "suck juices from the stems of plants and gnaw holes in roots, stalks, buds, ears, and pods. They attack pasture grasses, cereal and forage crops, young corn, nursery stock, and fruit trees. Fire ants may attack and kill newborn pigs, calves, sheep, and other animals; newly hatched chicks; and the young of ground-nesting birds." Auburn University entomologist F. S. Arant warned the ARS's F. J. Mulhern that other publications picked up "this misinformation" and spread it as fact. In 1958, Arant, Kirby L. Hayes, and Dan W. Speake had published an article proving that fire ants most often ate "house fly larvae, boll weevil grubs, cut-worms, and many other destructive insects in the field." Plant damage, they determined, was rare. "No damage to livestock has been observed," Arant insisted. "Cattle and sheep graze over the mounds and even lie down near them. Newly born livestock is rarely if ever killed. Continuing research demonstrated that fire ants at times did attack birds, young animals, and plants."[3] The ARS staff, of course, intentionally spread ghastly fire ant fables, attempting to alarm citizens and thus smooth the way for the mirex program.

In his 1968 presidential campaign, Richard Nixon came out for fire ant control, and, as Nixon appointees settled in at the USDA in 1969, fire ant advocates intensified their lobbying. Jim Buck Ross, Mississippi's commissioner of agriculture, who had run on a platform advocating fire ant eradication, worked closely with USDA undersecretary J. Phil Campbell, who had formerly been the Georgia agriculture commissioner. Ross shamelessly requested that the fire ant budget be increased from $25,000 to $41 million. He complained to Secretary of Agriculture Clifford M. Hardin that "our State is tired of living with Fire Ants. . . . In addition to destroying crops, machinery, and young livestock, this imported pest is a health hazard to the people and depress land value wherever it occurs."[4] In building support for a southern eradication campaign using mirex, Ross coordinated meetings, funded studies, and encouraged scientific pronouncements. He deplored the Johnson administration's lack of support for his endeavors but was optimistic that "new dynamic leaders," such as ARS deputy administrator Frank Mulhern, "offered *real* hope for a program." Ross called a well-publicized meeting of politicians, lobby groups, scientists, commissioners of agriculture, and the Southern Plant Board for September 8, 1969, in Montgomery. It was a love feast.[5]

After heptachlor and other chlorinated hydrocarbons had toxified the rural South without stunting fire ants, mirex emerged as the weapon of choice. In 1965 an ARS report claimed that mirex "now has universal acceptance," although wildlife bureaus had reservations about it, and some banned the substance. By April 1970 research revealed that mirex was especially lethal to crabs and shrimp. Despite these warnings, J. Phil Campbell simply called for more research. "We feel it would be highly irresponsible," he judged, "to attempt to take action on the basis of available information."[6]

Not all scientists agreed with Campbell's "spray now and do research later" methodology. On May 28, 1970, Louis N. Wise, vice president of the office for agriculture and forestry at Mississippi State University, denounced the ARS's decision to spray for fire ants and objected to the way the decision was made, disagreeing with Mississippi's Department of Agriculture in the process. Many of Mississippi State's top scientists—"all highly respected level-headed, agriculturally-oriented scientists"—Wise wrote to

Enough chemicals to treat one acre for fire ants. The small vial contains one spoonful (1.7 grams) of mirex, which would be dissolved in soybean oil and mixed with 1.25 pounds of granulated corncob grits in the large container. *Mississippi, September 1972, photograph by Murray Lemmon, 16-0972A1285-11, NARA.*

the ARS's Ned D. Bayley, had raised questions about the program. They were particularly appalled by mirex's potential effect on fish. At the time, Mississippi had some 20,000 acres in catfish production, representing an investment of $20 million. Preliminary research indicated that catfish accumulated mirex. "There will be all manner of hell raised by the catfish farmers, and rightly so, if their product is condemned by the Pure Food and Drug people," Wise warned. He explained that "the effects of Mirex on shrimp, oysters and other crustaceans," while not completely understood, was worrisome. Given the minuscule economic impact of fire ants, he argued that threatening the state's catfish industry was foolish. "Although you are well aware of it, as a reminder I might add that Mirex is on the suspect list as a carcinogenic material," he cautioned. He berated the ARS approach of "let's get started and we will adjust if and when problems arise." Wise allowed that "this gives me the feeling the idea is to jump in and do the job before the significance or consequence of the act is realized." It was difficult for Mississippi State scientists to speak out, he explained, because "the fire ant issue is so emotionally and politically charged that this approach is not very effective." He asked for a meeting of state and federal scientists to evaluate the issue. "If we don't do this, the more wild-eyed, purist type environmentalists are going to discredit all pesticides," he warned. "In this connection, I want to point out that in the case of the fire ant eradication program we are laying our collective necks on the block for something which, as I have already indicated, most objective scientists agree is of limited economic consequence."[7]

Initial spraying results revealed that Wise had offered prudent advice. "It appears safe to conclude that Mirex applied for control of the imported fire ant," L. D. Newsom reported in June 1970, "is solely responsible for the substantial residues of this insecticide that have been found in a large number of species of wildlife." He demanded that the fire ant campaign be halted and was alarmed that it had gone so far. The fire ant was not of economic importance, he fumed, nor had eradication been demonstrated anywhere with mirex or any other chemical. No studies had been completed on nontarget organisms. Newsom reviewed findings both on how mirex killed fish and how mirex residues accumulated in wildlife. Applying mirex "to such a large area of the United States, much of which is already overburdened with organochlorine insecticide residues, would be scientifically indefensible," he proclaimed.[8] Such questions should have been at the top of the ARS research agenda. Unfortunately, the agency ignored the prescient critiques of F. S. Arant, Louis Wise, and L. D. Newsom. It was not science that drove ARS decisions, but rather its own compulsion to please state agricultural leaders, lobbyist groups, political backers, and, of course, chemical companies.

Newsom had warned the ARS about poorly researched chemical application during the earlier fire ant campaign. In Louisiana, for example, there had been an increase in sugarcane borer infestations and a corresponding decrease in predators after heptachlor application. The fire ant, Newsom advised, was an important predator of sugarcane borers. In the treated area, rice stink bugs and leaf hoppers increased while crayfish in rice fields were all killed.[9] George W. Irving Jr. dismissed Newsom's warnings. Mirex applied at a rate of 1.7 grams per acre, he claimed, would not endanger wildlife. Since mirex was a chlorinated hydrocarbon, he callously reasoned, "it was to be expected that there would be some accumulation in the fatty tissues of some nontarget organisms." Ground-feeding insects picked up mirex, he admitted, and there was also toxic accumulation in birds.[10] This disturbing evidence did not halt the control program. ARS administrators downplayed the extent of fire ant control, arguing that compared to the amount of chemicals used by farmers in 1970, mirex would have an insignificant impact on the environment.

Disregarding contrary evidence, ARS representatives bragged to the Southern Plant Board in July 1970 that a three-year study "has shown that the fire ant can be eradicated, and any technical problems that exist, or are contemplated, can be overcome." The total projected cost came to "more than $200 million." In late August 1970, the ARS's T. C. Byerly perused "information on Mirex relevant to its alleged carcinogenicity" but set aside a study that indicated mirex caused cancer in mice. "In our opinion, the evidence is quite insufficient to establish as fact that Mirex is a carcinogen," he wrote. Byerly worried more about an approaching July 27 thirty-minute ABC news program on mirex and the fire ant campaign. "It is obvious that this attention at this time to this very sensitive problem is unlikely to be helpful to us," he fretted. Indeed, the general public was far more environmentally aware than it had been in the 1950s, and opposition grew so strong that it succeeded in temporarily halting the project. On August 5, the Environmental Defense Fund (EDF) sued to enjoin the USDA from continuing the mirex campaign. "Mirex residues become increasingly concentrated as they move up the food chain, thereby reaching their highest concentrations in fish, birds and man at the top of these food chains," it warned. In April 1971 a federal court denied the EDF motion to halt the program. T. C. Byerly continued shuffling health reports and fire ant program criticisms, all the while insisting that health threats to wildlife or humans were "trivial."[11]

In large measure due to *Silent Spring,* environmental issues were now of concern to many Americans; Clarence Cottam was no longer a soloist singing of ARS perfidy. In October 1970, Agnes B. Royce of Coral Gables, Florida, expressed incredulity that the USDA "would dump such vast amounts of poison" on fire ants. She worried about its effects on wildlife and about children being harmed. A month later Mrs. Robert E. C. Weaver of New Orleans wrote to Congressman Hale Boggs asking why "cancer-causing Mirex" would be used against fire ants since it killed fish and wildlife, interfered with reproduction in chickens, and persisted in the soil. She advocated a tactical mound-by-mound attack rather than indiscriminate aerial spraying. Finally, she insisted that property owners had the right to protect their land from "broadcast application of this poison without . . .

consent." Mrs. John L. Heard Jr. of New Orleans complained of the $200 million price tag. "If you will recall," she wrote to Secretary of Agriculture Clifford Hardin in November, "a similar attempt in the late 1950's using heptachlor and dieldrin did not succeed in eradicating the fire ant. It did succeed in killing [a] large number of fowl, fish, reptiles, and animals." The Earth Committee of Spring Hill College in Mobile, Alabama, sent the USDA a petition against the fire ant program with 394 signatures.[12]

Letters complaining of the neo–fire ant campaign reflected not only environmental awareness but also an impatience with cliched justifications for massive insect-control projects. In September 1970, Michael Owen Willson registered his displeasure that the USDA planned "to use 450 million pounds of mirex on fire-ant colonies in 14 Southern states." He lamented that "your department is as intent on destroying all other animals in the 14 states as it is on destroying the ants. Bureaucratic stupidity such as this must cease if we are to leave anything alive on this planet for the enjoyment of future generations." Alice Bailey, who lived in Fayetteville, Arkansas, learned that mirex was toxic to fish, wildlife, and possibly humans. "Since Mirex is more persistent than DDT, is a possible carcinogen, and concentrates in the food chain," she wrote in October 1970, "why is your department spending millions of dollars for widespread spraying in nine southeastern states?" Jerry and Sally Nagel of Johnson City, Tennessee, reminded Senator Howard Baker in November 1970 that mirex was dangerous and that the earlier fire ant program was flawed, writing, "An earlier effort in the 1950's was a complete failure with many disastrous side effects on wildlife and cattle." In November, biologist Theron J. Liskey wrote from Bridgewater, Virginia, to Senator Harry F. Byrd that the USDA and chemical companies "are again leagued in a $200 million boondoggle to 'eradicate' the *fire ant* in nine Southeastern states where it was 'eradicated' in the late 1950's at a horrible cost to tax-payers, the environment and to wildlife." He speculated that people's health over a wide area "likely suffered because of this irresponsible, senseless, asinine action." Liskey wanted "this 12-year planned attack on the United States stopped." But even as public criticism mounted, the USDA's N. P. Raiston assured Senator Thomas R. Eagleton, "No instance has come to our attention where the insecticide mirex as used in the imported fire ant program has killed fish, wildlife, domestic animals, or birds."[13]

Meanwhile, Jim Buck Ross discovered that an environmental group in Starkville had filed a lawsuit on August 5 to halt the fire ant campaign.

"These groups acting under the umbrella of protecting the environment," he fretted, "pose a serious threat to American agriculture, not only to the fire ant program, but to the use of all of our agricultural chemicals." He insisted that the fire ant "is a major pollutant of the environment—much more so than the small amount of mirex bait that is required to eradicate it."[14]

Attempting to counter environmental protests, in late 1970 advocates of fire ant eradication increased the political pressure. A delegation visited J. Phil Campbell on December 14 to discuss the fire ant campaign, and the strategy developed in this meeting later appeared in a letter. In January, fire ant supporters would flood the Nixon White House with ten thousand letters, while Campbell would make supportive telephone calls to southern agriculture commissioners. Campbell would also encourage letters to congressional delegations. A Republican National Committee member suggested that Campbell contact Clarke Reed, a powerful Mississippi Republican, who "could be most helpful in keeping the program."[15] The alliance of lobbyists, extension agents, chemical companies, and state farm leaders, eradication backers hoped, would generate a positive public response that would increase support for the program.

Without scientific or even practical reasons to spray mirex, proponents resorted to political pressure generated through local, state, and federal supporters. As planned, fire ant supporters organized a letter-writing campaign and relied upon the federal extension service to manage their effort. The county agent in Scott County, Mississippi, for example, distributed instructions on a mimeographed sheet: "If you are *for* a FIRE ANT PROGRAM then write a short letter to the President; address below, and get it to the County Agent's Office by January 13. The ones against this program are already writing him." Scott County needed 1,000 letters, he announced, urging, "Just two or three sentences, including, 'I am for the Fire Ant ERADICATION PROGRAM AND will appreciate you supporting this program.'" One resident simply wrote across the mimeographed form, "I'm for fire ant program" and sent it in. Jessie C. Haley wrote from Harriston, Mississippi, "The Imported Fire Ant is the most vicious pest we have ever had in our county. They are destroying our land, our game and some of our fish. They are harmful to the human being."[16]

This lobbying effort did not go unopposed, for environmentalists countered with strong letters condemning the program. Many southerners recalled the first fire ant campaign, dwelling on how chemicals operated in the environment. Elizabeth Brockman from St. Petersburg, Florida, wrote to President Nixon in December 1970 and praised his remarks on the environment in his State of the Union message. She warned Nixon of the planned fire ant campaign, continuing, "I remember specifically that about fifteen years ago, when malathion was sprayed from airplanes to eliminate a variety of fruit-fly that had invaded Florida, so many birds were killed that hordes of enormous green cabbage worms appeared and ate practically all the leaves off poinsettias and other shrubs." Mrs. Roger W. Brooking of Tucker, Georgia, painted a gloomy scenario for Nixon. Mirex was unsuccessful in eradicating fire ants in Georgia, and research had revealed carcinogenic properties that should have removed it from the market. "I've hesitated starting a family because of our 'tomorrow,'" she wrote. "I feel betrayed and helpless against the march of progress with its accompanying waste and pollution that is engulfing us all." She detested greed and felt that her vote did not accomplish anything. "We need help and encouragement from those with the influence—instead of a feeling that we're being manipulated and 'sold out' for the $," she pleaded.[17]

From Decatur, Georgia, Evelyn Angeletti presented eloquent and informed arguments. "Mirix, like aldrin, dieldrin and heptachlor, produces malignant tumors in mice," she informed Nixon in December, conceding that its effect on humans was uncertain. "Even the most ingenious layman," she argued about aerial spraying, "could not devise a method better calculated to contact larger numbers of nontarget organisms and fewer fire ant mounds." Instead of "unscientific, indiscriminate, and haphazard" aerial spraying that would cost $200 million, she suggested applying mirex directly to nesting sites. "Individual citizens like me," she lamented, "have no lobby in Washington as does the pesticide industry."[18]

Many letters recommended common sense. Writing from Gainesville, Florida, Martha F. Fielder speculated that the fire ant eradication program was born of "ignorance, and lack of far-sightedness for possible ecological results." Alternatively, she thought it had its roots in some "vested industry" (she mentioned the livestock industry) or "through some political payoff to the Mirex company." She predicted a disaster in the shrimping industry. "If the Mirex company needs financial support," she suggested, "I would rather you *give* them the $200,000,000.00 free and clear—but keep that pesticide out of the South and let us keep our fire ants, crayfish, and shrimp." Donald G. Legg of Notasulga, Alabama, reported to the USDA that it was all people could do to keep fire ants "out of the house, out of your food, and out of your bed." Still, he argued, "I would rather share my house, food, and bed with the ants than to have the U.S.D.A. attempt to exterminate them with pesticides." Nora E. Fogle from Menifee, Arkansas, informed Nixon that farmers used common sense when dealing with fire ants. "Just as small fires can be put out by one person (even a female)," she wrote, "so can individual ant hills be destroyed by pouring fuel oil or gasoline down the hole of the ant hill."[19]

Since *Silent Spring,* the press had increased its environmental coverage. On December 13, 1970, *New York Times* reporter Jon Nordheimer revealed that the mirex program would last a dozen years, spray 130 million acres with mirex, cost $200 million, and "eradicate the fire ant, which causes discomfort to picknickers and agricultural field workers." Already, Nordheimer pointed out, 14 million acres had been sprayed, including 600,000 acres near Tampa, Florida, where a state official pronounced it "devastating" to crustaceans in the Gulf of Mexico. Mirex was a suspect in blue crab kills near Savannah, where 2 million acres were sprayed. In April 1971, Dr. Charles F. Wurster, professor of environmental science at SUNY–Stony Brook, labeled the program "a political pork barrel that helps pump money into the states." The *New York Times* reported that the Allied Chemical Company manufactured mirex for the fire ant campaign and that—perhaps not so coincidentally—its factory was located in Aberdeen, Mississippi.[20]

The pro and con letters registered not only with the White House and the bureaucracy but also with congressmen. As chairman of the Agriculture Appropriations Subcommittee, Mississippi representative Jamie Whitten exerted enormous control over the budget. Even after pesticide control moved to the EPA, Whitten maintained budget oversight. In both fire ant campaigns, he championed eradication. It is unknown how much the fifteen Allied Chemical Company workers in Aberdeen, Mississippi, and the million dollars in mirex sales colored his enthusiasm. He continued to rage at *Silent Spring* and denounce Rachel Carson at every opportunity. Despite the volume of letters that protested the second war on fire ants and increasing reports of collateral damage, the program continued. In March 1971, EPA head William D. Ruckelshaus cancelled all federal registrations of mirex, but manufacturer Allied Chemical Company appealed the decision. In May 1972, the EPA narrowed the window for mirex application; as part of a compromise plan, Allied agreed to monitor mirex levels in aquatic areas for two years. Despite serious environmental problems, the ARS sprayed 20 million acres with mirex in 1972.[21]

Gradually, state funds for fire ant control dried up, and in 1974 the USDA announced that it would abandon the mirex fire ant eradication plan. The EPA's rules limiting the use of mirex, the USDA charged, made the program "unworkable." The state of Mississippi then bought the Allied plant and continued the state program, but new evidence showed undisputedly that mirex was a carcinogen, thus dooming its efforts. Since 1958, federal and state programs had cost $148 million without diminishing the number of fire ants. The ARS's J. Phil Campbell stubbornly insisted that eradication was "environmentally sound." Harvard University fire ant expert, Dr. Edward O. Wilson, disagreed. "The fire ant control program in the South," he judged, "is the Vietnam of entomology." He doubted the wisdom of the entire misadventure. "Ants," he pointed out, "occupy one of the middle links in the food chain—they consume other small insects and are scavengers that recycle nutrients from decomposing material in fields and woodlands—and no one has ever examined the long-range consequences of killing off lots of ants."[22]

The second war against fire ants ended much as the first had done. Ignoring warnings about dangers to fish and wildlife, the ARS sprayed mirex as recklessly as it had sprayed heptachlor, dieldren, and aldrin in the first campaign. Fire ants responded by prolific breeding, resistance, and expanding to new ranges. Although environmentalists wrote numerous letters opposing the fire ant program, they could not prevent it. In both fire ant projects, the ARS contrived to garner support from Congress, agribusiness lobbyists, and the federal extension service. Combined with other insect-control projects and general agricultural use, pesticides left a toxic legacy of illness, wildlife destruction, and contaminated factory sites and dumps.

The complexity of agricultural chemicals, the power of advertising to mask the dangers of toxins, and misplaced trust in federal regulatory agencies engendered public apathy about pesticides and other potentially harmful products. While the environmental movement slowed the chemical treadmill, the struggle over regulation and safety continues. The rapid spread of synthetic pesticides after World War II embodied claims that scientific intervention would ensure better health and wholesome food. To a large extent this claim was fulfilled. Still, evidence of pesticide residues, unlabeled genetically modified foods, unease over mad cow disease, and disturbing reports of continuing pesticide dangers alarm consumers. A history of misuse, mislabeling, and misleading statements has undermined trust in regulatory agencies. Just as swirling chlorinated hydrocarbon residues from a half-century ago endanger Arctic Circle human and animal health, dead zones in the Gulf of Mexico and in the Chesapeake Bay illustrate the enormous collateral damage inflicted by chemical manufacturing, agricultural pesticides runoff, and waste from confined animal production. No model can accurately predict the impact that genetically engineered fish, animals, and plants will have on their natural counterparts—or, for that matter, on human health and the environment. What amounts to the premature adoption of new products has been a constant since at least World War II. The corporate compulsion to market first, test later, and resist regulation has left a legacy of widespread sickness and death. Corporate lobbyists continually refine tools that manipulate politicians, and agribusiness inexorably has its way with regulators. Environmentalists gain a few steps with one administration and lose them to the next.[23]

The pesticide wreckage generated in the quarter century after World War II is thus a cautionary history. In an era that began with Hiroshima and progressed through nuclear testing, radiation experiments, ARS control projects, the marketing of poorly tested and inadequately labeled pesticides, Vietnam defoliation with Agent Orange, corporate pesticide dumping, and the creation of toxic waste zones, it is not difficult to imagine chemical companies and their political and bureaucratic allies giving a higher priority to the death of insects than to the health of humans.

NOTES

NOTES TO CHAPTER 1

1. Testimony of Dr. A. E. Wood, 223, 225, 234, *Charles Lawler* v. *W. T. Skelton* et al. (no. 7868), 241 Miss. 274 (1960), in trial transcript, Mississippi Department of Archives and History, Jackson, Miss. Hereafter cited as *Charles Lawler* v. *W. T. Skelton* et al. (no. 7868), 241 Miss. 274 (1960).

2. Edmund Russell, *War and Nature: Fighting Humans and Insects with Chemicals from World War I to "Silent Spring"* (Cambridge, Eng., 2001), 86, 146–49, 161–65; D. F. Murphy to Watson Davis, May 9, 1944, box 262, folder 7, Science Service Records (RU 7091), Smithsonian Institution Archives, Washington, D.C. Marcel LaFollette called my attention to the Science Service Records. The fight against agricultural pests profited from intense research during World War I to discover countermeasures to chemical warfare, which ultimately killed some 90,000 people. Following the war, the Chemical Warfare Service and the U.S. Department of Agriculture's Bureau of Entomology further blurred the boundaries between human and insect research. By elevating insects as its new enemy, the Chemical Warfare Service continued its research, understanding that lethal insecticides were potentially harmful to humans. See Edmund P. Russell, "'Speaking of Annihilation': Mobilizing for War against Human and Insect Enemies, 1914–1945," *Journal of American History* 82 (March 1996): 1505–29. For a broad treatment of synthetic chemicals, see Thomas R. Dunlap, *DDT: Scientists, Citizens, and Public Policy* (Princeton, 1981); Robert L. Rudd, *Pesticides and the Living Landscape* (Madison, 1964); Frank Graham Jr., *Since "Silent Spring"* (Boston, 1970). On the development of synthetic chemicals and early resistance research, see John S. Cecatti, "Biology in the Chemical Industry: Scientific Approaches to the Problem of Insecticide Resistance, 1920s–1960s," *Ambix* 51 (July 2004): 135–47.

3. USDA assistant secretary Charles F. Brannan claimed that "a great deal of experimental work has been done" with DDT and that "it is safe to recommend the use of certain types of DDT insecticides for the control of certain insect pests." See Charles F. Brannan to Edwin Arthur Hall, October 11, 1945, chemicals, General Correspondence, 1906–75, Records of the Secretary of Agriculture, Record Group 16, National Archives and Records Administration (hereafter cited as GC, 1906–75, SOA, RG 16, NARA). For a USDA statement on synthetic pesticides, see C. V. Bowen and S. A. Hall, "The Organic Insecticides," *Yearbook of Agriculture, 1952* (Washington, D.C., 1952), 209–17. On the development of DDT and early health studies, see Russell, *War and Nature,* 145–64.

On plant growth regulators, see John W. Mitchell, "Plant Grown Regulators," *Science in Farming: The Yearbook of Agriculture, 1943–1947* (Washington, D.C., 1947), 256–66.

4. Marla Cone, "Of Polar Bears and Pollution," *Los Angeles Times,* June 19, 2003, A1, A8–9; DeNeen L. Brown, "Poisons from Afar Threaten Arctic Mothers, Traditions," *Washington Post,* April 11, 2004, A25.

5. "Better Farming through Research," Shell Agricultural Laboratory brochure, 1946, box 298, folder 8, Science Service Records (RU 7091), Smithsonian Institution Archives, Washington, D.C.

6. Frank Thone to Atherton Richards, March 4, 1943, and Richards to Thone, July 12, 1943, box 252, folder 5; Thone to Richards, August 7, 1947, and Richards to Thone, August 13, 1947, box 290, folder 10, ibid.; interview with the Reverend John Harris, Franklin, La., May 28, 1988, by Lu Ann Jones, Oral History of Southern Agriculture, Archives Center, National Museum of American History, Washington, D.C.

7. *Science in Farming; Yearbook of Agriculture: Insects* (Washington, D.C., 1952); Albert P. Brodell, Paul E. Strickler, and Harold C. Phillips, "Extent and Cost of Spraying and Dusting on Farms, 1952," Statical Bulletin No. 156, USDA, Agricultural Research Service, April 1955. On biological control, see Richard C. Sawyer, *To Make a Spotless Orange: Biological Control in California* (Ames, 1996).

8. Christopher J. Bosso, *Pesticides and Politics: The Life Cycle of a Public Issue* (Pittsburg, 1987), 47–53; H. L. Haller and Ruth L. Busbey, "The Chemistry of DDT," *Science in Farming,* 616–22.

9. Bosso, *Pesticides and Politics,* 53–60, 71–78; Louis Pyenson, "Pesticides Are Surer and Safer This Year," *New York Times,* January 29, 1956, 136.

10. Wheeler McMillen, *Bugs or People?* (New York, 1965), 87; Russell, *War and Nature,* 165–83.

11. Bernard E. Conley, "Incidence of Injury with Pesticides," *Journal of the American Medical Association* (hereafter cited as *JAMA*) 163 (April 13, 1957): 1338–40. See James Whorton, *Before "Silent Spring": Pesticides and Public Health in Pre-DDT America* (Princeton, 1974); Scott R. Baker and Chris F. Wilkinson, eds., *The Effects of Pesticides on Human Health* (Princeton, 1990); James E. Davies and R. Doon, "Human Health Effects of Pesticides," in *"Silent Spring" Revisited,* ed. Gino J. Marco, Robert M. Hollingsworth, and William Durham (Washington, D.C., 1987), 113–24; Robert K. Plumb, "Dichloro-diphenyl-trichloroethane," *New York Times,* January 16, 1955, SM38.

12. W. E. Westlake to S. A. Hall, December 15, 1960, pesticide chemicals, box 5, Entomology Research Division, Director's Correspondence, 1959–65, Records of the Agricultural Research Service, Record Group 310, National Archives and Records Administration. Hereafter cited as ERD, DC, 1959–65, ARS, RG 310, NARA.

13. On the emerging law, see George C. Chapman, "Crop Dusting—Scope of Liability and a Need for Reform in the Texas Law," *Texas Law Review* 40 (April 1962): 527–41; "Crop Dusting: Legal Problems in a New Industry," *Stanford Law Review* 6 (December 1953): 69–90; "Liability for Chemical Damage from Aerial Crop Dusting," *Minnesota Law Review* 43 (January 1959): 531–44; L. S. Carsey and J. S. Covington Jr., "Legal Responsibilities Arising from Aerial Application of Agricultural Chemicals," *Insurance Counsel Journal* 30 (July 1963): 417–22; "Regulation and Liability in the Application

of Pesticides," *Iowa Law Review* 49 (Fall 1963): 135–49; Jay A. Sigler, "Controlling the Use of Pesticides," *Journal of Public Law* 15 (1966): 311–23; "Agricultural Pesticides: The Need for Improved Control Legislation," *Minnesota Law Review* 52 (June 1968): 1242–60; N. William Hines, "Agriculture: The Unseen Foe in the War on Pollution," *Cornell Law Review* 55 (May 1970): 740–60; Robert Van Den Bosch, "Insecticides and the Law," *Hastings Law Journal* 22 (February 1971): 615–28; C. Gray Burdick, "Liability in Crop Dusting: A Survey," *Mississippi Law Journal* 42 (Winter 1971): 104–16; Randy L. Harshman, "Liability for Pesticide Application," *Gonzaga Law Review* 9 (Spring 1974): 816–25; J. Steve Massoni, "Environmental Law: Agricultural Pesticides," *Washburn Law Journal* 13 (Winter 1974): 53–67; Ellen S. Greenstone, "Farmworkers in Jeopardy: OSHA, EPA, and the Pesticide Hazard," *Ecology Law Quarterly* 5 (1975): 69–121; Charles P. Cockerill, "Agricultural Pesticides: The Urgent Need for Harmonization of International Regulation," *California Western International Law Journal* 9 (Winter 1979): 111–38. Also see *Kentucky Aerospray* v. *Mays,* 251 S.W.2d, 460–62 (Ky., 1952); and *L. W. Wall* v. *Wendell Trogdon,* 107 S.E.2d, 757 (N.C., 1959).

14. "Toxicity of DDT for Man," *JAMA* 135 (December 6, 1947): 939; Theodore T. Stone and Lee Gladstone, "DDT," *JAMA* 145 (April 28, 1951): 1342; Nathan J. Smith, "Clinical Notes, Suggestions, and New Instruments," *JAMA* 136 (February 14, 1948): 469–71; John F. Marchand, "Microtests for Cholinesterase," *JAMA* 149 (June 21, 1952): 738–40; "Committee on Pesticides," *JAMA* 150 (November 1, 1952): 903–4; "Aldrin and Dieldrin Poisoning," *JAMA* 146 (May 26, 1951): 378–79.

15. "Parathion Poisoning in Citrus Grove Operations," *JAMA* 152 (July 11, 1953): 1071; Jere W. Annis and John W. Williams, "Change in Electrolytes in a Case of Parathion Poisoning," *JAMA* 152 (June 13, 1953): 594–96; Archer S. Gordon and Charles W. Frye, "Large Doses of Atropine," *JAMA* 159 (November 19, 1955): 1181–84; Ernest M. Dixon, "Dilatation of the Pupils in Parathion Poisoning," *JAMA* 163 (February 9, 1957): 444–45.

16. "Committee on Pesticides," *JAMA* 156 (November 6, 1954): 969.

17. "Committee on Pesticides," *JAMA* 168 (October 25, 1958): 1061.

18. Rev. Gilbert P. Herrman to Dwight D. Eisenhower, June 18, 1953; W. L. Popham to C. E. Schoenhals, July 16, 1953; Byron T. Shaw to Herrman, July 18, 1953; Herrman to Shaw, July 23, 1953; M. R. Clarkson to Herrman, August 13, 1953, agricultural research, insecticides, box 891, ARS, RG 310, NARA.

19. On milk and meat residues, see Pete Daniel, *Lost Revolutions: The South in the 1950s* (Chapel Hill, 2000), 71–75.

20. Unidentified Science Service draft, n.d., box 298, folder 8, Science Service Records (RU 7091), Smithsonian Institution Archives, Washington, D.C.

NOTES TO CHAPTER 2

1. Testimony of Charles Lawler, in trial transcript, 256–63; "Declaration," 2–11, *Charles Lawler* v. *W. T. Skelton* et al. (no. 7868), 241 Miss. 274 (1960).

2. Testimony of Charles Lawler, in trial transcript, 256–70; "Declaration," 2–11, *Charles Lawler* v. *W. T. Skelton* et al. (no. 7868), 241 Miss. 274 (1960); statement of

C. E. Lawler, June 28, 1958, and statement of Irene Lawler, June 28, 1958, in Townsend/Hulen files, *Lawler* v. *Skelton,* Townsend, McWilliams, and Holladay Office, Drew, Miss. Malathion was toxic enough to damage paint on automobiles. See *Motors Insurance Corporation* v. *Aviation Specialties, Inc.* v. *The United States of America,* 304 F. Supp. 973 [1969 U.S. Dist LEXIS 10677 (Michigan 1969)].

3. See Agency for Toxic Substances and Disease Registry, "Endrin" (http://www.atsdr.cdc.gov/toxprofiles/phs89.html), "Malathion" (http://www.atsdr.cdc.gov/MHMI/mmg154.html), "Xylenes" (http://www.atsdr.cdc.gov/tfacts71.html).

4. Testimony of Charles Lawler, in trial transcript, 270–73, *Charles Lawler* v. *W. T. Skelton* et al. (no. 7868), 241 Miss. 274 (1960); interview with Mr. C. E. Lawler, October 6, 1958, in Townsend/Hulen files, *Lawler* v. *Skelton,* Townsend, McWilliams, and Holladay Office, Drew, Miss.

5. Testimony of Dr. Mary Elizabeth Hogan, 739–41, in trial transcript, *Charles Lawler* v. *W. T. Skelton* et al. (no. 7868), 241 Miss. 274 (1960); "memorandum of conference with Dr. Hogan," n.d., mentioned in Elizabeth Hulen to Pascol Townsend, August 15, 1958, in Townsend/Hulen files, *Lawler* v. *Skelton,* Townsend, McWilliams, and Holladay Office, Drew, Miss.; interview with Dr. Mary Elizabeth Hogan, Helen Neal, and Linda Henderson, by Pete Daniel, April 1, 2003, Pearl, Miss.

6. "Health Department Says Crop Poisoning OK If with Caution," August 15, 1957, *Delta Democrat-Times,* 1; "Glen Allan Malady," August 16, 1957, *Delta Democrat-Times,* 4; "Take Blood Samples in Glen Allan Areas," August 20, 1957, *Delta Democrat-Times,* 1; "Suspect Asiatic Flu in Washington and Sharkey Counties," August 21, 1957, *Delta Democrat-Times,* 1; "Suspected Flu Confined to Small Areas in State," August 23, 1957, *Delta Democrat-Times,* 8.

7. M. R. Clarkson to E. D. Burgess, October 26, 1956, and M. R. Clarkson, memoranda for the files, September 10, 1957, regulatory crops 1, plant pest control division, ARS, RG 310, NARA; memorandum of conference with Mr. Charles Lawler, August 9, 1958; memorandum of conference with Dr. Hogan (ca. August 15, 1958); interview with Mr. C. E. Lawler, October 6, 1958; undated note from Irene Lawler, in Townsend/Hulen files, Townsend, McWilliams, and Holladay Office, Drew, Miss; interview with Dr. Mary Elizabeth Hogan, Helen Neal, and Linda Henderson. On Hayes, see Dunlap, *DDT,* 70–71. On Rachel Carson's regard for Hayes, see Linda Lear, *Rachel Carson: Witness for Nature* (New York, 1997), 334–35.

8. Interview with Dr. Mary Elizabeth Hogan, Helen Neal, and Linda Henderson.

9. Walter Sillers to A. L. Gray, September 23, 1963, box 19, folder 28, Walter Sillers Papers, Delta State University Archives, Cleveland, Miss.

10. A. L. Gray to Walter Sillers, September 26, 1963, ibid.

11. Testimony of Charles Lawler, in trial transcript, 270–73, *Charles Lawler* v. *W. T. Skelton* et al. (no. 7868), 241 Miss. 274; Pascol Townsend to Elizabeth Hulen, July 30, 1958; memorandum of conferences with Charles Lawler, July 30, August 9, 1958; interview with Mr. C. E. Lawler, October 6, 1958; Irene Lawler to Elizabeth Hulen, March 19, 1962, in Townsend/Hulen files, *Lawler* v. *Skelton,* Townsend, McWilliams, and Holladay Office, Drew, Miss.

12. Interview with John McWilliams and Lawson Holladay, by Pete Daniel, June 27, 2002, Drew, Miss.

13. Elizabeth Hulen to Pascol Townsend, August 15, 29, September 4, 29, 1958, March 19, 1959, January 29, February 15, 1960; Hulen to Dr. Lee R. Reid, October 11, 15, 1958; Townsend and Hulen notes for final argument, in Townsend/Hulen files, *Lawler* v. *Skelton,* Townsend, McWilliams, and Holladay Office, Drew, Miss.

14. Memorandum of conference with Charles Lawler, July 30, August 9, 1958; Elizabeth Hulen to Pascol Townsend, August 15, 1958; Hulen to Townsend, August 29, 1958; Hulen to Dr. W. R. Webb, September 3, 1958; Hulen to Townsend, September 29, October 28, 1958; Townsend to Eddie Peacock, February 13, 1959, in ibid.

15. E. F. Knipling, office memorandum, March 11, 1960, cotton insects research branch, box 3, ERD, DC, 1959–65, ARS, RG 310, NARA.

16. Interview with Judge Charlotte Buchanan, by Pete Daniel, June 27, 2002, Indianola, Miss.

17. Testimony of W. T. Skelton, in trial transcript, 144–203, *Charles Lawler* v. *W. T. Skelton* et al. (no. 7868), 241 Miss. 274 (1960).

18. Testimony of V. A. Johnson, in trial transcript, 509–20, ibid.

19. Testimony of J. L. Turk, in trial transcript, 563–69, ibid.

20. Ibid., 570.

21. Testimony of John P. Martin, in trial transcript, 575–91, ibid. On nozzle settings, see James E. Garton, "A Graphic Solution of Airplane Sprayer Problems," Oklahoma Agricultural Experiment Station Miscellaneous Publication No. MP-20, May 1951.

22. Testimony of Charles E. Lawler, in trial transcript, 246–57, 292, *Charles Lawler* v. *W. T. Skelton* et al. (no. 7868), 241 Miss. 274 (1960).

23. Ibid.

24. Ibid., 258–61.

25. Ibid., 262–69.

26. Testimony of Charles S. Lawler, in trial transcript, 217–21.

27. Testimony of Johnnie B. McCaleb, in trial transcript, 458–74, ibid.; trial notes of Townsend and Hulen, in Townsend/Hulen files, *Lawler* v. *Skelton,* Townsend, McWilliams, and Holladay Office, Drew, Miss.

28. Testimony of Dr. Mitchell Zavon, in trial transcript, 350–52, *Charles Lawler* v. *W. T. Skelton* et al. (no. 7868), 241 Miss. 274 (1960); Jaques Cattell, ed., *American Men of Science: A Biographical Directory* (Tempe, 1961), 4562.

29. Testimony of Dr. Mitchell Zavon, in trial transcript, 350–68, *Charles Lawler* v. *W. T. Skelton* et al. (no. 7868), 241 Miss. 274 (1960). See R. I. Krieger and T. M. Dinoff, "Malathion Deposition, Metabolite Clearance, and Cholinesterase Status of Date Dusters and Harvesters in California," *Archives of Environmental Contamination and Toxicology* 38 (2000): 546–53.

30. Testimony of Dr. Mitchell Zavon, in trial transcript, 366–79, *Charles Lawler* v. *W. T. Skelton* et al. (no. 7868), 241 Miss. 274 (1960).

31. Lear, *Rachel Carson,* 334; Bob Wyrick, "EPA Officials Devised Cancer Tests on People," *Washington Post,* June 23, 1977, A1.

32. Wayland J. Hayes Jr., "Agricultural Chemicals and Public Health," *Public Health Reports* 69 (October 1954): 893–98; Hayes, "Diagnostic Problems in Toxicology (Agriculture)," *AMA Archives of Environmental Health* 3 (July–December 1961): 55–56.

33. Testimony of Dr. Mitchell Zavon, in trial transcript, 380–95, *Charles Lawler* v. *W. T. Skelton* et al. (no. 7868), 241 Miss. 274 (1960).

34. Ibid., 395–401.

35. Ibid., 411–20.

36. Testimony of Griffith E. Quinby, 552–53, ibid.; Cattell, ed., *American Men of Science.* 3271.

37. Testimony of Griffith E. Quinby, in trial transcript, 552–55, *Charles Lawler* v. *W. T. Skelton* et al. (no. 7868), 241 Miss. 274 (1960). See William M. Upholt, Griffith E. Quinby, Gordon S. Batchelor, and James P. Thompson, "Visual Effects Accompanying TEPP Induced Miosis," *AMA Archives of Ophthalmology* 56 (July 1956): 128–34; Griffith E. Quinby and Allen B. Lemmon, "Parathion Residues as a Cause of Poisoning in Crop Workers," *JAMA* 166 (February 15, 1958): 740–46. Quinby also studied antidotes for parathion poisoning. See Griffith E. Quinby and Gordon B. Blappison, "Parathion Poisoning: A Near-Fatal Pediatric Case Treated with 2-Pyridine Aldoxime Methiodide (2-PAM)," *AMA Archives of Environmental Health* 3 (July–December 1961): 52–56.

38. Testimony of Griffith E. Quinby, in trial transcript, 556–58, *Charles Lawler* v. *W. T. Skelton* et al. (no. 7868), 241 Miss. 274. On cholinesterase testing, see "Parathion Poisoning in Citrus Grove Operations," *JAMA* 152 (July 11, 1953): 1071; Tatsuji Namba, Martin Greenfield, and David Grob, "Malathion Poisoning: A Fatal Case with Cardiac Manifestations," *Archives of Environmental Health* 21 (October 1970): 533–39.

39. Testimony of Griffith E. Quinby, in trial transcript, 467, 561–62, *Charles Lawler* v. *W. T. Skelton* et al. (no. 7868), 241 Miss. 274 (1960). See Griffith E. Quinby, Kenneth C. Walker, and William F. Durham, "Public Health Hazards Involved in the Use of Organic Phosphorus Insecticides in Cotton Culture in the Delta Area of Mississippi," *Journal of Economic Entomology* 51, no. 6 (December 1958): 831–38, exhibit P-22, in trial transcript, *Charles Lawler* v. *W. T. Skelton* et al. (no. 7868), 241 Miss. 274 (1960); Richard L. Fowler, "Manifestations of Cottonfield Insecticides in the Mississippi Delta," *Journal of Agricultural and Food Chemistry* 1 (June 10, 1953): 469–73.

40. Testimony of Dr. A. A. Aden, in trial transcript, 608–12; testimony of Dr. Rozelle Hahn, in trial transcript, 323–42, *Charles Lawler* v. *W. T. Skelton* et al. (no. 7868), 241 Miss. 274 (1960).

41. Testimony of Dr. A. A. Aden, in trial transcript, 608–30, ibid.; trial notes of Townsend and Hulen, in Townsend/Hulen files, *Lawler* v. *Skelton,* Townsend, McWilliams, and Holladay Office, Drew, Miss.

42. Watts R. Webb to A. A. Aden, June 1, 1957, exhibit A at 631A; testimony of Dr. A. A. Aden, in trial transcript, 633–34, *Charles Lawler* v. *W. T. Skelton* et al. (no. 7868), 241 Miss. 274 (1960).

43. Testimony of Cecil Black, 705–715; testimony of L. L. Vance, 716–19; testimony of Frank L. Tindall, 720–26; testimony of B. G. Ames, 727–38, ibid.

44. Testimony of Dr. Mary Elizabeth Hogan, 739–47; Quinby, Walker, and Durham, "Public Health Hazards Involved in the Use of Organic Phosphorus Insecticides," 831–38, ibid.

45. Quinby, Walker, Durham, "Public Health Hazards Involved in the Use of Organic Phosphorus Insecticides," 831–35, ibid.

46. Ibid., 835.

47. Ibid., 835–38.

48. Appellant brief, *Lawler* v. *Skelton;* brief on behalf of W. T. Skelton and V. A. Johnson (no. 41,855), in Townsend/Hulen files, *Lawler* v. *Skelton,* Townsend, McWilliams, and Holladay Office, Drew, Miss.

49. *Lawler* v. *Skelton,* 130 So.2d, 565–70.

50. B. F. Smith to Boswell Stevens, July 17, 1961, and Stevens to Smith, July 20, 1961, subject files, 1959–73, Boswell Stevens Papers, Special Collections Department, Mitchell Memorial Library, Mississippi State University, Mississippi State; Porter W. Peteet et al., "Petition to Appear as *Amicus Curiae* and for Leave to File Brief on Suggestion of Error," in trial transcript, *Charles Lawler* v. *W. T. Skelton* et al. (no. 7868), 241 Miss. 274 (1960).

51. "Insecticide Recommendations of the Entomology Branch for the Control of Insects Attacking Crops and Livestock, 1956 Season," Agriculture Handbook No. 103, ARS and Federal Extension Service, USDA, March 1956, 84–87; "Insecticide Recommendations of the Entomology Branch for the Control of Insects Attacking Crops and Livestock, 1959 Season," Agriculture Handbook No. 120, ARS and Federal Extension Service, USDA, February 1959, 109–12.

52. "Suggestion of Error of Appellee, W. T. Skelton," in trial transcript, *Charles Lawler* v. *W. T. Skelton* et al. (no. 7868), 241 Miss. 274 (1960).

53. B. F. Smith to Boswell Stevens, July 17, 1961; Stevens to Smith, July 20, 1961, subject files, 1959–73, Boswell Stevens Papers, Special Collections Department, Mitchell Memorial Library, Mississippi State University, Mississippi State.

54. Pascol Townsend to Charles Lawler, October 17, 1961; Elizabeth Hulen to Hon. M. M. McGowan, October 30, 1961; Forrest Cooper to Pascol Townsend, February 10, 23, 1962; Elizabeth Hulen to Farish, Keady, & Campbell, March 1, 1962, in Townsend/Hulen files, *Lawler* v. *Skelton,* Townsend, McWilliams, and Holladay Office, Drew, Miss.

55. Mrs. Charles Lawler to Elizabeth Hulen, March 19, 1962; Hulen to Townsend, March 20, 1962; Petition for Approval of Settlement, *Lawler* v. *Skelton,* et al., Circuit Court of Jefferson Davis County; Hulen to Townsend, May 9, 1962; Townsend to Hulen, May 16, 1962, in ibid.

NOTES TO CHAPTER 3

1. Mabry I. Anderson, *Low and Slow: An Insider's History of Agricultural Aviation* (Perry, Ga., 1986), 3–40; Eldon W. Downs and George F. Lemmer, "Origins of Aerial Crop Dusting," *Agricultural History* 39 (July 1965): 123–35; Corley McDarment, "The Use of Airplanes to Destroy the Boll Weevil," *McClure's Magazine* 57 (August 1924): 90–102.

2. Interview with Cotton Carnahan, Chuck Thresto, and Jack Shannon, Clarksdale, Miss., October 8, 1987, by Lu Ann Jones, Oral History of Southern Agriculture, National Museum of American History, Washington, D.C.; Anderson, *Low and Slow,* 121. On post–World War II developments in spray apparatus, see J. S. Yuill, D. A. Isler, and George D. Childress, "Research on Aerial Spraying," *Yearbook of Agriculture: Insects* (Washington, D.C., 1952), 252–58.

3. Interview with Cotton Carnahan, Chuck Thresto, and Jack Shannon.

4. A. J. Loveland to W. Lee O'Daniel, October 11, 1948; Charles F. Brannan to J. W. Fulbright, June 21, 1948; Brannan to Wayne Morse, July 9, 1948, chemicals, GC, 1906–75, SOA, RG 16, NARA; Joseph C. Chamberlin, Charles W. Getzendaner, Harold H. Hessig, and V. D. Young, *Studies of Airplane Spray-Deposit Patterns at Low Flight Levels,* Technical Bulletin No. 1110 (Washington, D.C., 1955). For cases concerning drift, see *Chapman Chemical Company* v. *Taylor,* 222 S.W.2d, 820–28 (Ark., 1949); *Burns* v. *Vaughn,* 224 S.W.2d, 365–66 (Ark., 1949); *Gotreaux* v. *Gary,* 94 So.2d, 293–95 (La., 1957); *Heeb* v. *Prysock,* 245 S.W.2d, 577–80 (Ark., 1952); *Alexander* v. *Seaboard Air Line R. Co.,* 71 S.E.2d, 299–305 (S.C., 1952); Ned D. Bayley to Charles A. Mosher, July 11, 1969, box 5080, pesticides, GC, 1906–75, RG 16, NARA.

5. Interview with Cotton Carnahan, Chuck Thresto, and Jack Shannon.

6. Ibid.; Anderson, *Low and Slow,* 73–74, 114–15.

7. Interview with R. C. Colvin, Greenwood, Miss., October 22, 1987, by Lu Ann Jones, Oral History of Southern Agriculture, National Museum of American History, Washington, D.C.

8. Testimony of John P. Martin, 586; testimony of G. G. Ames, 729; testimony of Cecil Black, 707, in trial transcript, *Charles Lawler* v. *W. T. Skelton* et al. (no. 7868), 241 Miss. 274 (1960); "Flagman in Tall Cotton Struck by Sprayer Plane," *Delta Democrat-Times,* September 5, 1957, 1.

9. Interview with Cotton Carnahan, Chuck Thresto, and Jack Shannon; Donovan Webster, "A Reporter at Large: Heart of the Delta," *New Yorker,* July 8, 1991, 49–52.

10. Webster, "A Reporter at Large," 56–58.

11. Interview with Cotton Carnahan, Chuck Thresto, and Jack Shannon. Although there were some federal regulations regarding crop dusting, most states had laws that controlled the application of pesticides. See Ned D. Bayley to Charles A. Mosher, July 11, 1969, pesticides, box 5080, GC, 1906–75, SOA, RG 16, NARA; Anderson, *Low and Slow,* 44–48.

12. Webster, "A Reporter at Large," 48.

13. Ibid., 49–52; interview with Cotton Carnahan, Chuck Thresto, and Jack Shannon.

14. Neal Landy to Joseph S. Clark Jr., May 27, 1957; Landy to True D. Morse, May 27, 1957; Roger Hinds to Alexander H. Smith, July 27, 1957, insects, GC, 1906–75, SOA, RG 16, NARA. See also Richard Sidwell to B. B. Hickenlooper, August 21, 1956, insecticides, ibid.

15. V. E. Weyl to E. D. Burgess, May 8, 1957; L. F. Curl, memo on telephone conversation with Congresswoman Katherine St. George's office, May 9, 1957, regulatory crops, box 752, ARS, RG 310, NARA.

16. G. W. Irving Jr. to Orville Freeman, June 20, 1967, and attached, "Imported Fire Ant: Sequence of Events," insects, box 4679, GC, 1906–75, SOA, RG 16, NARA; Bosso, *Pesticides and Politics,* 79–108.

17. "Wildlife Service Gets 2 Aides to Director," *Washington Post,* April 24, 1953, 17; John B. Oakes, "Conservation: Political Issue," *New York Times,* May 3, 1953, X28; "Former Chief of Agency Drops to 4th," *Washington Post,* July 23, 1954, 26; "Clarence Cottam of Texas Refuge," *New York Times,* April 3, 1974, 46; "Clarence Cottam, Wildlife Director," *Washington Post,* April 6, 1974, B3; Lear, *Rachel Carson,* 257.

18. Clarence Cottam to Miller F. Shurtleff, October 14, 1958; Cottam to Shurtleff, November 19, 1958, insects, GC, 1906–75, SOA, RG 16, NARA.

19. Miller Shurtleff to Ezra Taft Benson, November 21, 1958, and attachment; Shurtleff to Clarence Cottam, December 3, 1958, and attached anonymous to Cottam, December 15, 1958; Cottam to Shurtleff, December 19, 1958, insects, GC, 1906–75, SOA, RG 16, NARA.

20. Clarence Cottam to Miller Shurtleff, February 15, 1960; Cottam to Shurtleff, October 27, 1960; E. D. Burgess to James C. Davis, December 8, 1959, insecticides, ibid. Emphasis in original.

21. Miller Shurtleff to Charlie [?], March 9, 1960, insects, ibid.

22. Bosso, *Pesticides and Politics,* 113–14. On the larger questions raised by the fire ant control programs, see Joshua Blu Buhs, *The Fire Ant Wars: Nature, Science, and Public Policy in Twentieth-Century America* (Chicago, 2004).

23. W. H. Welch Jr. to Orville L. Freeman, February 13, 1961; Howard W. Johnson to Welch, February 23, 1961, chemicals, box 6, ERD, DC, 1959–65, ARS, RG 310, NARA.

24. *S. L. Winston, Jr. et al.* v. *State of Louisiana, DOT,* 352 So.2d ,752 (1977) [1977 La. App. LEXIS 5245 (Louisiana)].

25. C. S. Baker to Lyndon Baines Johnson, October 6, 1960, insects, box 3455; J. V. Beck to John D. Dingell, August 21, 1961; Dingell to Orville Freeman, August 29, 1961; Frank J. Welch to Dingell, September 7, 1961, insects, box 3610, GC, 1906–75, SOA, RG 16, NARA.

26. R. D. Radeleff to A. M. Lee, May 7, 1962; Radeleff, "Investigation of Cattle Losses Near Monroe, Louisiana," chemicals, box 10, ERD, DC, 1959–65, ARS, RG 310, NARA.

27. "Remarks of R. D. Radeleff, Animal Disease and Parasite Research Division, during Discussion Meeting for Pesticide Panel of Federal Council of Science and Technology," November 6, 1962, chemicals, box 11, ibid. Emphasis in original.

28. Mrs. Thomas R. Dillon to Senator Paul H. Douglas, February 25, 1960, box 3455; Dillon to Elmer J. Hoffman, April 20, 1960, box 3456; Mrs. Arthur M. Jens Jr. to O. L. Meyer, April 28, 1960, box 3455, GC, 1906–1975, SOA, RG 16, NARA; Adam Rome, "'Give Earth A Chance': The Environmental Movement and the Sixties," *Journal of American History* 90 (September 2003): 534–41.

29. Statement of Ruth Graham Desmond, U.S. Senate, Subcommittee on Reorganization and International Organizations of the Committee on Government Operations, *Interagency Coordination in Environmental Hazards (Pesticides): Hearing on S. Res. 288,* 88th Cong., 2d sess., April 15, 1964, 2030.

30. O. D. Brattan to Estes Kefauver, April 25, 1961, insects, box 3610; John R. Priora to Estes Kefauver, April 23, 1961, and attached, Paul Farleigh, "Game Chief Denounces Poison Spray Program," clipping from *Memphis Press-Simitar;* Frank J. Welch to Kefauver, May 15, 1961, insects, box 3456, GC, 1906–75, SOA, RG 16, NARA.

31. Alice Durand to Orville Freeman, February 13, 1962, insecticides, box 3782; Press Release, National Association of County Agricultural Agents, February 18, 1960, insects, public relations 2, box 228, ibid.

32. Barbara Bolling to Kenneth Birkhead, August 4, 1962; Herb Ansel to Bolling, n.d., insecticides, box 3782, ibid.

33. Kenneth M. Birkhead to Barbara Bolling, September 24, 1962, ibid.

34. Lear, *Rachel Carson,* 7–243.

35. Ibid., 306, 312–13.

36. Ibid., 429, 448–49; Jack Gould, "TV: Controversy over Pesticide Danger Weighed," *New York Times,* April 4, 1963, 95.

37. B. T. Shaw to Orville Freeman, July 11, 1962, insects, box 3782, GC, 1906–75, SOA, RG 16, NARA.

38. Rodney E. Leonard to Orville Freeman, July 12, 1962, ibid.; Howard W. Johnson, "Comments on Rachel Carson's Articles in the *New Yorker,*" September 10, 1962; A. M. Lee to Howard W. Johnson, August 30, 1962, committees, chemicals, box 10, ERD, DC, 1959–65, ARS, RG 310, NARA. For favorable and unfavorable reactions to *Silent Spring,* see Lear, *Rachel Carson,* 396–456.

39. William C. Parrish to Estes Kefauver, January 20, 1963; Orville Freeman to Parrish, February 5, 1963, insecticides, box 3959, GC, 1906–75, SOA, RG 16, NARA.

40. Damon Stetson, "Specialist Cites Insecticide Peril," *New York Times,* March 4, 1963, 7.

41. Neal Stanford, "Parley to Sift Theories on Farm-Pest Controls," *Christian Science Monitor,* February 2, 1966, 7.

NOTES TO CHAPTER 4

1. T. C. Byerly to Orville Freeman, February 28, 1963; Byron T. Shaw to Jerome B. Wiesner, March 1, 1963; Freeman to Wiesner, May 1, 1963, and attached part 3, "The Hazards of Using Pesticides," insecticides, box 3959, GC, 1906–75, SOA, RG 16, NARA; Bosso, *Pesticides and Politics,* 120–25.

2. Statement of Jerome B. Wiesner, "Use of Pesticides: A Report of the President's Science Advisory Committee," U.S. Senate, Committee on Government Operations, Subcommittee on Reorganization and International Organizations, Coordination of Activities Relating to the Use of Pesticides, *Hearing on S Res. 27,* 88th Cong., 1st sess., May 16, 1963, 33–61; Zuoyue Wang, "Responding to *Silent Spring:* Scientists, Popular Science Communication, and Environmental Policy in the Kennedy Years," *Science Communication* 19 (December 1997): 141–63.

3. Testimony of Orville Freeman, U.S. Senate, Committee on Government Operations, Subcommittee on Reorganization and International Organizations, Coordination of Activities Relating to the Use of Pesticides, *Hearing on S Res. 27,* 88th Cong., 1st sess., May 16, 1963, 84–88.

4. Ibid., June 4, 1963, 205–6.

5. Statement of Rachel Carson, ibid., 206–19.

6. Testimony of Rachel Carson, ibid., 219–20.

7. Statement of Parke C. Brinkley, June 25, 1963, ibid., 251–65; Howard Hass to George Barnes, June 25, 1963, insecticides, box 3959, GC, 1906–75, SOA, RG 16, NARA.

8. Testimony of Mitchell Zavon, U.S. Senate, Committee on Government Operations, Subcommittee on Reorganization and International Organizations, Coordination of Activities Relating to the Use of Pesticides, *Hearing on S Res. 27,* 88th Cong., 1st sess.,

June 25, 1963, 274–75, 294–95, 304–13; Howard Hass to George Barnes, June 25, 1963, insecticides, box 3959, GC, 1906–75, SOA, RG 16, NARA; Robert C. Toth, "Pesticide Makers Score U.S. Report," *New York Times,* June 26, 1963, 40.

9. Kenneth F. Maxcy to the Surgeon General, May 1, 1945; Lowell E. Noland to the Surgeon General, October 15, 1946; Qualifications Record; Hayes memo to Surgeon General, PHS, November 27, 1967; David J. Spencer to Chief, Office of Personnel, February 12, 1964, and attachment, "Justification for Nomination of Wayland J. Hayes Jr., Medical Director, for Distinguished Service Medal," personnel papers, Wayland J. Hayes Jr., Office of the Historian, Public Health Service.

10. Testimony of Wayland J. Hayes Jr., U.S. Senate, Committee on Government Operations, Subcommittee on Reorganization and International Organizations, Coordination of Activities Relating to the Use of Pesticides, *Hearing on S Res. 27,* 88th Cong., 1st sess., July 17, 1963, 407–12, 451–52.

11. Testimony and statement of Dr. Malcolm M. Hargraves, ibid., 484–85, 487, 490–92, 497, 511–12; Lear, *Rachel Carson,* 320, 332; Robert C. Toth, "Doctors Cautious on Pesticide Role," *New York Times,* July 18, 1963, 42.

12. Testimony of Joseph H. Holmes, U.S. Senate, Committee on Government Operations, Subcommittee on Reorganization and International Organizations, Coordination of Activities Relating to the Use of Pesticides, *Hearing on S Res. 27,* 88th Cong., 1st sess., June 25, 1963, 526–34.

13. Statement of Dr. W. C. Hueper, July 18, 1963, ibid., 704–9, 714; Lear, *Rachel Carson,* 332, 356.

14. Warren G. Magnuson to Orville L. Freeman, August 12, 1963; Freeman to Magnuson, August 27, 1963, insecticides, box 3959, GC, 1906–75, SOA, RG 16, NARA.

15. George Barnes to Orville Freeman, August 14, 1963, ibid.; Freeman to Nyle C. Brady, April 17, 1964, insecticides, box 4132, ibid.

16. N. C. Brady to Howard Bertsch, T. C. Byerly, Lloyd H. Davis, and B. T. Shaw, August 6, 1964, insecticides, box 4132, ibid.

17. Kenneth M. Lynch to H. F. Kraybill, October 12, 1964; Kraybill to Lynch, October 20, 1964; Lynch to Kraybill, November 3, 1964; Kraybill to Lynch, November 12, 1964; H. W. Johnson to E. R. Goode, November 25, 1964, chemicals, box 17, ERD, DC, 1959–65, ARS, RG 310, NARA.

18. Clarence Cottam to Orville Freeman, May 4, 1965, insects, box 4309, GC, 1906–75, SOA, RG 16, NARA. See also Eloise W. Kailin to Editors of Government Executive, August 27, 1969, pesticides, box 5081, ibid., and enclosed article, Eloise W. Kailin and Alicia Hastings, "Electromyographic Evidence of DDT-Induced Myasthenia," *Medical Annals of the District of Columbia* 35 (May 1966): 237ff.

19. Clarence Cottam to N. C. Brady, June 2, 1965; Brady to Cottam, May 21, 1965, insects, box 4309, GC, 1906–75, SOA, RG 16, NARA; "DDT Found Harmless in Human Body," *Washington Post,* December 30, 1955, 8; "U.S. Aide Says DDT Caused No Illness," *New York Times,* February 22, 1958, 6; "Pesticide Deaths Put Below One in a Million," *New York Times,* November 16, 1962, 10.

20. B. F. Smith to Delta Council Board of Directors and Agricultural Committee Members, November 2, 1962, and enclosure, "The Desolate Year," Early C. Ewing Jr.

records, Delta and Pine Land Company Records, Special Collections, Mitchell Memorial Library, Mississippi State University; Jamie L. Whitten to George Mehren, November 22, 1967, chemicals, box 2, Office of the Administrator, Central Correspondence File, 1967–73, ARS, RG 310, NARA (hereafter cited as OA, CCF, 1967–73, ARS, RG 310, NARA).

21. Nick Kotz and Morton Mintz, "Pesticide Firms Aided Whitten Book," *Washington Post,* March 14, 1971, 1; Jamie L. Whitten to George Mehren, November 22, 1967, pesticides, box 4713, GC, 1905–76, SOA, RG 16, NARA.

NOTES TO CHAPTER 5

1. Graham, *Since "Silent Spring,"* 96–103; Robert J. Anderson, "The Significance of the Mississippi River Fish Kills," November 30, 1964, chemicals, box 17, ERD, DC, 1959–65, ARS, RG 310, NARA; U.S. Senate, Hearings, Committee on Government Operations, Subcommittee on Reorganization and International Organizations, *Interagency Coordination in Environmental Hazards (Pesticides),* 88th Cong., 2nd sess., April 7, 1964, exhibit 158, "Pesticide Caused Fish Kills as Reported by State Agencies to United States Public Health Service, 1960–1963," 1671–77.

2. David Anderson, "Poisons Kill Fish in the Mississippi," *New York Times,* March 22, 1964, 79; Donald Janson, "Pesticides Fatal to Gulf Shrimp," *New York Times,* March 26, 1964, 26; Janson, "Bayous Despair at Fish Scourge," *New York Times,* March 29, 1964, 60.

3. Jean M. White, "Fish Kills in Mississippi Stir New Pesticide Probe on Hill," *Washington Post,* March 27, 1964, A8; "More 'Silent Spring,'" *Washington Post,* March 30, 1964, A16.

4. Summary, "Meeting on the Subject of Fish-Kill in the Lower Mississippi River," Taft Sanitary Engineering Center, Cincinnati, Ohio, April 1, 1964; Nyle C. Brady to Orville Freeman, April 4, 1964, insecticides, box 4132, GC, 1906–75, SOA, RG 16, NARA.

5. Nyle C. Brady to Orville Freeman, April 3, 1964, insecticides, box 4132, GC, 1906–75, SOA, RG 16, NARA; Walter Sullivan, "Pesticide Fatal to Fish Is Traced," *New York Times,* April 5, 1964, 70; John W. Finney, "U.S. Agency Warned Year Ago That Pesticides Could Kill Fish," *New York Times,* April 7, 1964, 55.

6. Testimony of James M. Quigley, 1681–82; testimony of Donald I. Mount, 1683; testimony of James M. Hundley, 1685; testimony of Donald McKernan, 2003; exhibit 160, PHS, "Report on Investigation of Fish Kills in Lower Mississippi River, Atchafalaya River, and Gulf of Mexico," U.S. Department of Health, Education, and Welfare, April 6, 1964, 1698, 1706, 1712; all in Senate Committee, *Interagency Coordination in Environmental Hazards (Pesticides).*

7. "Report on Investigation of Fish Kills," Senate Committee, *Interagency Coordination in Environmental Hazards (Pesticides)* 1725, 1756, 1759.

8. Jean M. White, "Fish Kill Laid to Endrin, Maker Disputes Verdict," *Washington Post,* April 10, 1964, A2; John W. Finney, "Pesticide Maker Challenges U.S.," *New York Times,* April 10, 1964, 27; "Use of Pesticides Backed by Maker," *New York Times,* April 11, 1964, 6; Neal Stanford, "Fish Loss Speeds U.S. Pesticide Probe," *Christian Science*

Monitor, April 11, 1964, 7; "For Action on Pesticides," *New York Times,* April 13, 1964, 28; *Washington Post,* April 13, 1964, A16.

9. Louis Silver, "Pesticides Stoutly Defended against Fish-Kill Charges," *Memphis Commercial Appeal,* April 17, 1964, 4.

10. John W. Finney, "Freeman Admits Pesticide Snarl," *New York Times,* April 16, 1964, 61; "Fish Kill Laid to Plant Waste," *Washington Post,* April 16, 1964, A25; testimony of Orville Freeman, Senate Committee, *Interagency Coordination in Environmental Hazards (Pesticides),* 2005–7, 2014; statement of Orville Freeman, ibid., 2016.

11. Testimony of Orville Freeman, Senate Committee, *Interagency Coordination in Environmental Hazards (Pesticides),* 2022–25. See also Bosso, *Pesticides and Politics,* 129–30.

12. Orville Freeman, memorandum for files, April 15, 1964, insecticides, box 4132, GC, 1906–75, SOA, RG 16, NARA.

13. Thomas L. Kimball to Orville L. Freeman, May 1, 1964, ibid.

14. Orville Freeman to Nyle C. Brady, April 17, 1964; Freeman to Brady, June 16, 1964, ibid.

15. Julian Huxley to the *New York Times,* April 19, 1964, E8.

16. "Preliminary Report on Investigation of Industrial Pesticide Plants in the Mississippi Delta," April 24, 1964, insecticides, box 4132, GC, 1906–75, SOA, RG 16, NARA; "Summary of Water Pollution Survey Activities by the U.S. Public Health Service in the Vicinity of Memphis, Tenn., 1964–65," Senate Committee, *Interagency Coordination in Environmental Hazards (Pesticides),* 1791–92; "Shaping Memphis Environment," *Memphis Commercial Appeal,* April 22, 1990, Velsicol Chemical Corporation File, Memphis Room, Memphis Public Library.

17. Nyle C. Brady to Orville Freeman, April 27, 1964, and attached, "Preliminary Report on Investigation of Industrial Pesticide Plants in the Mississippi Delta," April 24, 1964, insecticides, box 4132, GC, 1906–75, SOA, RG 16, NARA; Stewart Udall, "Our Seesaw Battle against a Silent Spring," *True Magazine* (August 1965): 72, copy with W. C. Shaw to Brady, July 30, 1965, pesticides, box 4359, GC, 1906–75, SOA, RG 16, NARA; "Pollution Stirs Memphis Dispute," *New York Times,* January 17, 1965, 47; "Velsicol Mulls $25,000 City Problem," *Memphis Commercial Appeal,* July 23, 1965; "Velsicol Agrees To Clean Sewer of Endrin Waste," *Memphis Commercial Appeal,* August 3, 1965; "Velsicol To Get OK on Sewer Cleanup," *Memphis Commercial Appeal,* September 14, 1965. Last three citations in Velsicol Chemical Corporation File, Memphis Room, Memphis Public Library.

18. B. T. Shaw, through Nyle C. Brady, to Orville Freeman, May 14, 1964, insecticides, box 4132, GC, 1906–75, SOA, RG 16, NARA.

19. Brown Alan Flynn, "Velsicol Installs Treatment Plant," *Memphis Press Scimitar,* June 5, 1963; "Chlorine Damage Involves Two Major Manufacturers," *Memphis Commercial Appeal,* June 13, 1963; "Accord Reached in Velsicol Suits," *Memphis Commercial Appeal,* October 17, 1966; "$153,125 Awarded in Velsicol Suit," *Memphis Commercial Appeal,* March 28, 1967, all in Velsicol Chemical Corporation file, Memphis Room, Memphis Public Library; "Pollution Stirs Memphis Dispute," *New York Times,* January 17, 1965, 47.

20. Walter Sullivan, "Pesticide Maker Tied to Fish Kill," *New York Times,* April 23, 1964, 41; John Herbers, "Memphis May Act against Pesticide," *New York Times,* April 26, 1964, 60; John W. Finney, "U.S. Finds Endrin Kills River Fish," *New York Times,* May 6, 1964, 52; John W. Finney, "Fish-Kill Theory of U.S. Disputed," *New York Times,* May 7, 1964, 34; "Senate Study Hinted on U.S. Indictment of Pesticide in River," *New York Times,* June 17, 1964, 38.

21. "Ribicoff Plans Law for Plant Inspection in Fish-Kill Issue," *Washington Post,* April 24, 1964, A9; "Pesticide Data Leaked to Public," *Christian Science Monitor,* April 25, 1964, 4; R. J. Anderson, memorandums for the files, July 8, July 10, July 16, 1964, insecticides, box 4132, GC, 1906–75, SOA, RG 16, NARA.

22. "Fish Deaths Laid to Endrin by U.S.," *New York Times,* June 27, 1964, 9; John W. Finney, "Fish-Kill Theory Is Altered Again," *New York Times,* June 30, 1964, 21; Paul A. Schuette, "Officials Questioned on Pesticide Endrin," *Washington Post,* June 30, 1964, B12; Neal Stanford, "Endrin Report Fuels Debate," *Christian Science Monitor,* July 1, 1964, 1; "Company Assails U.S. Endrin Tests," *New York Times,* July 29, 1964, 68.

23. Morris Cunningham, "EPA Claims Action at Chemical Dumps," *Memphis Commercial Appeal,* October 31, 1975; "Velsicol Says Waste Site Probably Affected Water," *Memphis Commercial Appeal,* October 7, 1978; Tom Stone, "Velsicol Agrees with Ban Order on Dumping," *Memphis Press-Scimitar,* March 2, 1972; "Shaping Memphis Environment," *Memphis Commercial Appeal,* April 22, 1990, all in Velsicol Chemical Corporation File, Memphis Room, Memphis Public Library. On Velsicol's problems with environmental violations, see "Velsicol Asks Delay on Ban," *Memphis Commercial Appeal,* August 7, 1975; "Judge Refuses To Dismiss Suit against Velsicol," *Memphis Press-Scimitar,* November 29, 1977; Kathrine Barrett and Dorothy Bland, "Ill Workers' Tests Find Velsicol Link," *Memphis Commercial Appeal,* July 9, 1978; "Sludge Clearance Being Discussed," *Memphis Commercial Appeal,* May 19, 1978, all in Velsicol Chemical Corporation File, Memphis Room, Memphis Public Library; Tom Charlier, "Cypress Soil to be Studied," April 28, 2003, *GoMemphis* (http://www.gomemphis.com); Tom Charlier, "Study to Focus on Cancer Link to Creek," *Memphis Commercial Appeal,* June 10, 2004; "Study of Cypress Creek Toxicity Worthy," *Memphis Commercial Appeal,* June 16, 2004. On Velsicol's veracity regarding tests on heptachlor, see *U.S.* v. *Harvey S. Gold, et al.* 470 F. Supp. 1336 [1979 U.S. Dist. LEXIS 12906 (1979)]; Craig E. Colten, "The Big Kill: Endrin in the Mississippi and the Search for a New Sink," *Bulletin of the Illinois Geographical Society* 43 (Spring 2001): 30–37. For analyses of corporate pollution south of Baton Rouge, Louisiana, see also Barbara L. Allen, *Uneasy Alchemy: Citizens and Experts in Louisiana's Chemical Corridor Disputes* (Cambridge, Mass., 2003), and Craig E. Colten, ed., *Transforming New Orleans and Its Environs: Centuries of Change* (Pittsburgh, 2000).

24. W. F. Barthel to D. R. Shepherd, April 14, 1967, regulatory and control, box 16; "Visit with Mr. Jim Ewart of Velsicol Chemical Corporation, Memphis, Tennessee," May 10, 1967, regulatory and control, box 14, OA, CCF, 1967–73, ARS, RG 310, NARA.

25. W. F. Barthel to D. R. Shepherd, April 13, 1967 regulatory and control, box 16, ibid.

26. "Visit with Mr. Green of Niagara Chemical Company, Greenville, Mississippi," May 4, 1967; R. J. Anderson to S. H. Bear, July 24, 1967, regulatory and control, box 15, ibid.

27. W. J. Anthony to Robert J. Anderson, July 5, 1967; S. H. Bear to Anderson, June 27, 1967, regulatory control, box 14, ibid.

28. Interview with John McWilliams and Lawson Holladay.

29. *A. R. Boroughs* v. *Leo Joiner,* 337 So.2d, 340 [1976 Ala. LEXIS 1645 (Alabama)]; *George R. Green* v. *John Zimmerman and R. L. McNeil,* 269 S.C. 535; 238 S.E.2d, 323 [S.C. LEXIS 333 (South Carolina)].

30. L. Blus, E. Cromartie, L. McNease, and T. Joanen, "Brown Pelican: Population Status, Reproductive Success, and Organochlorine Residues in Louisiana, 1971–1976," *Bulletin of Environmental Contamination and Toxicology* 22 (1979): 128–35.

31. Roy Reed, "Pesticides Make Cotton Prosper But Endanger Life," *New York Times,* December 26, 1969, 26.

NOTES TO CHAPTER 6

1. Parke C. Brinkley to Charles S. Murphy, June 15, 1964, and attached clipping from *Agricultural Chemicals,* June 1964, insecticides, box 4132, GC, 1906–75, SOA, RG 16, NARA.

2. George Barnes to Orville Freeman, May 26, 1964; Barnes to Charles S. Murphy, June 17, 1964, ibid.

3. Ibid.

4. Harold Lewis to George Mehren, November 26, 1965, pesticides, box 4358, ibid.

5. J. A. Noone to Robert J. Anderson, July 20, 1967, meetings, box 10, OA, CCF, 1967–73, ARS, RG 310, NARA.

6. Donald Lerch to Clarence Palmby, February 4, 1969, pesticides, box 5080, GC, 1906–75, SOA, RG 16, NARA.

7. N. C. Brady to T. C. Byerly, B. T. Shaw, and V. L. Harper, December 10, 1964; Byerly to Brady, January 11, 1965, insects, box 4309, ibid.

8. Ned D. Bayley to Jamie Whitten, October 17, 1969, box 5081; Parke C. Brinkley to Clifford M. Hardin, May 27, 1970, box 5269, ibid.

9. Parke C. Brinkley to George W. Irving Jr., June 27, 1969, chemicals 1, box 25, OA, CCF, 1967–73, ARS, RG 310, NARA.

10. George W. Irving Jr. to Robert B. Rathbone, July 3, 1969; Marshall Gall to Irving, June 27, 1969; Irving to Parke C. Brinkley, July 11, 1969, ibid. Emphasis in original.

11. George W. Irving Jr. to Robert B. Rathbone, August 19, 1969; Rathbone to Irving, October 2, 1969; Irving to M. B. Green, October 7, 1969; D. L. Colinese to Irving, November 7, 1969, chemicals 1, box 41, ibid.

12. C. J. Robben to Robert Leggett, August 23, 1965; E. E. Saulmon to Leggett, September 24, 1965, pesticides, box 4358, GC, 1906–75, SOA, RG 16, NARA.

13. Jane Margaret and James V. Ward to Ralph Yarborough, June 26, 1969; Yarborough to Clifford M. Hardin, July 10, 1969; Mr. and Mrs. G. W. Gandy to Robert Dole, July 15, 1969, pesticides, box 5080, ibid. On drift, see *W. M. Young* v. *Reeford Darter,* 1961 Okla.

142; 363 P.2d, 829 [1961 Okla. LEXIS 390]; *V. L. Olmstead* v. *M. E. Reedy,* 1963 Okla. 268; 387 P.2d, 631 [1963 Okla. LEXIS 548]; *McKennon* v. *Jones,* 219 Ark. 671; 244 S.W.2d, 138 [1951Ark. LEXIS 588].

14. Roy J. Smith to John T. Holstun, July 27, 1967, chemicals 1, box 2; administrator ARS, memo to the Secretary, August 11, 1967, research farm 3, box 19, OA, CCF, 1967–73, ARS, RG 310, NARA.

15. Edward P. Cliff to the Secretary, July 11, 1968, pesticides, box 4851, GC, 1906–75, SOA, RG 16, NARA.

16. Bill Barksdale to Robert F. Eagen Jr., n.d.; Eagen to Barksdale, n.d.; Eagen to Kenneth C. Walker, January 4, 1967, research farm, box 19, OA, CCF, 1967–73, ARS, RG 310, NARA.

17. Stanley J. Medine to Clifford Hardin, July 19, 1969, chemicals 1, box 25, ibid.

18. Wayne M. Blickhahn to John Byrnes, June 24, 1968; Ned D. Bayley to Byrnes, July 29, 1968, pesticides, box 4851, GC, 1906–75, SOA, RG 16, NARA.

19. "Quantities of Pesticides Used by Farmers in 1966," Agricultural Economic Report No. 179, Economic Research Service, USDA, 1970, vi, 5, 10–11, 15–16, 19–21, 23–24; "Farmers' Use of Pesticides in 1971," Agricultural Economic Report No. 252, Economic Research Service, USDA, 1974, 4, 10–15; Census of Agriculture, 1997, table 1, "Historical Highlights: 1997 and Earlier Census Years," 10.

20. Leo G. K. Iverson to T. C. Byerly, May 4, 1971, chemicals, box 70, OA, CCF, 1967–73, ARS, RG 310, NARA.

21. Donald V. Eitzman and Sorrell L. Wolfson, "Acute Parathion Poisoning in Children," *American Journal of Diseases of Children* 114 (October 1967): 397–400, attached to Sam M. Gibbons to Orville L. Freeman, November 8, 1967, pesticides, box 4713; Herbert S. Harrison to G. G. Rohwer, August 25, 1970, and attached report, pesticides, box 5269, GC, 1906–75, SOA, RG 16, NARA.

22. Memorandum of telephone conversation between J. B. Corson and A. J. Dellavecchia, September 29, 1967, box 16, OA, CCF, 1967–73, ARS, RG 310, NARA; *Acady Farms Milling Company* v. *Betts,* 93 Ga. App. 255, 91 S.E.2d, 289 [1956 Ga. App. LEXIS 710]. See also Kirsten Waller, Thomas J. Prendergast, Alan Slagle, and Richard J. Jackson, "Seizures after Eating a Snack Food Contaminated with the Pesticide Endrin," *Western Journal of Medicine* 157 (December 1992): 648–51.

23. Herbert G. Lawson, "Death in the Fields," *Wall Street Journal,* July 16, 1971, 1, 16.

24. T. C. Byerly to Ned D. Bayley, September 2, 1969 (marked administratively confidential), pesticides, box 5081, GC, 1906–75, SOA, RG 16, NARA.

25. Kenneth C. Walker, "Report of Conference with Dr. Wayland J. Hayes, Jr., Vanderbilt University, Nashville, Tennessee, February 29, 1972," March 14, 1972, chemicals 1, box 91, OA, CCF, 1967–73, ARS, RG 310, NARA.

26. Bill Robinet, *By the Skin of My Teeth: A Cropduster's Story* (Veneta, Ore., 1997), 26–27, 41.

27. Robinet, *By the Skin of My Teeth,* 80, 236–37. See John A. Dellinger, "Monitoring the Chronic Effects of Anticholinesterase Pesticides in Aerial Applicators," *Veterinary and Human Toxicology* 27 (October 1985): 427–30.

28. Robinet, *By the Skin of My Teeth,* 68, 237–39; *Mississippi Valley Aircraft Service* v. *Brown,* 111 So.2d, 28 (Mississippi, 1959).

29. Ibid., 256–57, 262–65.

30. Ibid., 326–29, 332–34.

31. Ibid., 237–38.

32. Kenneth P. Cantor and Charles F. Booze, "Mortality among Aerial Pesticide Applicators and Flight Instructors," *Archives of Environmental Health* 45 (September–October 1990): 295–302.

33. Anne R. Yobs to T. C. Byerly, October 12, 1970, and attached, W. A. Williams, "The North Carolina State Board of Health Pesticides Project," July 31, 1970, pesticides, box 5268, GC, 1906–75, SOA, RG 16, NARA; Theodore R. Van Dellen, "How to Keep Well," *Washington Post*, July 15, 1967, C20.

34. Robert D. Dixon to J. Phil Campbell, November 7, 1969; Campbell to Dixon, December 4, 1969, pesticides, box 5081, GC, 1906–75, SOA, RG 16, NARA.

35. *Sanford and Elinor Kornberg* v. *Getz Exterminators*, 341 S.W.2d, 819 [1961 Mo. LEXIS 741]; *Wilbur Hampton* v. *Ruth Loper*, 402 S.W.2d 825 [1966 Mo. App. LEXIS 663].

36. *Gayle Laborde, et al.* v. *Velsicol Chemical Corporation*, 474 S.2d, 1320 [La. App. LEXIS 9555]; *Miller, Guardian* v. *Travelers Insurance Company*, 111 Ga. App. 245, 141 S.E.2d, 223 [1965 Ga. App. LEXIS 935].

37. U.S. House of Representatives, hearings before the Subcommittee on Energy and Commerce, "Controlling Airborne Releases of Chemicals," *Hearings on H.R. 2622*, 100th Cong., 1st sess., June 24, 1987, 11–21.

38. Testimony of Michelle Slowey, 97–101, ibid.

39. Interview with Juanita Russell, by Pete Daniel, June 30, 2002, Helena, Ark.

40. John Noble Wilford, "Deaths from DDT Successor Stir Concern," *New York Times*, August 21, 1970, 1, 15; "Safety Steps Taken on Deadly Pesticide," *New York Times*, December 1, 1970, 56.

41. John Noble Wilford, "U.S. Speeds up the Search for a Less Toxic Substitute for DDT," *New York Times*, August 27, 1970, 18.

42. Harry W. Hays to George W. Irving Jr., July 26, 1971, chemicals 1, box 70, OA, CCF, 1967–73, ARS, RG 310, NARA.

43. Ibid.

44. Ibid.; Waldemar Klassen to S. A. Hall, August 9, 1971, ibid.

45. H. Rex Thomas to T. W. Edminster, September 1, 1971 (draft); H. O. Graumann to T. W. Edminster, September 8, 1971, ibid.

46. C. H. Hoffman to Waldemar Klassen, August 24, 1971; W. C. Shaw to Klassen, August 26, 1971, ibid.; George W. Irving Jr. to Ned D. Bayley, July 13, 1970, pesticides, box 5269, GC, 1906–75, SOA, RG 16, NARA; Richard A. Lewis, "Cigaret Strife," *Wall Street Journal*, August 26, 1960, 1; "House Waits for Tobacco Control Clue," *Washington Post*, February 16, 1961, B3.

47. "Proposal Submitted by Entomology Research Division for Consideration of the Tobacco Research and Marketing Advisory Committee at its 1959–60 Meeting," committees, advisory, box 1, ERD, DC, 1959–65, ARS, RG 310, NARA; Sam R. Hoover to Ned D. Bayley, October 15, 1971, chemicals, box 70, OA, CCF, 1967–73, ARS, RG 310, NARA; "Registration of Endrin, An Insecticide, for Use on Tobacco is Dropped," *Wall Street Journal*, February 17, 1964, 9.

48. Harry W. Hays to Hiram Fong, November 21, 1967, regulatory and control 2, box 16, OA, CCF, 1967–73, ARS, RG 310, NARA.

49. W. C. Shaw to K. C. Walker, October 9, 1967, research farm 3, box 19, ibid.

50. Fred H. Husbands to R. J. Anderson, April 28, 1967; Kenneth C. Walker to Anderson, May 3, 1967, regulatory control, box 10; Kenneth C. Walker to Anderson, and attachment, "Pesticide Residues in Crops Grown in Rotation," May 10, 1967, chemicals 1, box 2; H. Wayne Bitting to John R. Hatchett, September 29, 1967, projects, box 13, OA, CCF, 1967–73, ARS, RG 310, NARA. For a statement of Shell Chemical Company and Velsicol Chemical Corporation's interpretation of such residues, see Bernard Lorant and M. J. Sloan to U.S. Department of Agriculture and Food and Drug Administration, June 21, 1967, box 16, ibid.

51. Melvin C. Tucker to J. Phil Campbell, July 25, 1969, chemicals 1, box 25, ibid.; Dave L. Pearce to Campbell, July 24, 1969, pesticides, box 5081, GC, 1906–75, SOA, RG 16, NARA; "Agriculture Dept. Suspends Use of DDT," *Washington Post,* July 10, 1969, A17.

52. K. R. Fitzsimmons to Clifford M. Hardin, December 10, 1969; Fitzsimmons to J. Phil Campbell, December 23, 1969; Ned D. Bayley to Fitzsimmons, February 10, 1970; George W. Irving Jr. to Bayley, February 25, 1970, pesticides, box 5070, GC, 1906–75, SOA, RG 16, NARA.

53. Robert White-Stevens to Ned D. Bayley, December 10, 1970; Bayley to White-Stevens, December 23, 1970, chemicals 1, box 47, OA, CCF, 1967–73, ARS, RG 310, NARA.

54. Edward M. Shulman to the Secretary, August 12, 1970, pesticides, box 5269, GC, 1906–75, SOA, RG 16, NARA.

55. Bosso, *Pesticides and Politics,* 139–42, 152–54.

NOTES TO CHAPTER 7

1. Statement of O. L. Kline, U.S. Senate, Subcommittee on Reorganization and International Organizations of the Committee on Government Operations, *Interagency Coordination in Environmental Hazards (Pesticides): Hearing on S. Res. 27,* 88th Cong., 2d sess., July 18, 1963, 790–91; U.S. House, Report, *Deficiencies in Administration of Federal Insecticide, Fungicide, and Rodenticide Act,* November 13, 1969, 91st Cong., 1st sess., 1969, H. Rept. No. 91–637, 3–4, 7, 11, 20–27, 33–36; U.S. House, Hearings before Subcommittee of the Committee on Government Operations, *Deficiencies in Administration of Federal Insecticide, Fungicide, and Rodenticide Act,* June 24, 1969, 91st Cong., 1st sess., May 7, 1969, 70; William M. Blair, "Pesticides Both Boon and Threat," *New York Times,* November 22, 1959, E8.

2. Statement of M. R. Clarkson, U.S. Senate Subcommittee, *Interagency Coordination in Environmental Hazards (Pesticides),* 737. On the importance of protest registration, see Graham, *Since "Silent Spring,"* 80.

3. Statement of O. L. Kline, U.S. Senate Subcommittee, *Interagency Coordination in Environmental Hazards (Pesticides),* , 790–91.

4. Thomas H. Harris to Harry W. Hays, June 19, 1967, regulatory and control 2, box 16, OA, CCF, 1967–73, ARS, RG 310, NARA.

5. Harry W. Hays to Thomas H. Harris, August 15, 1967, and appended notes, ibid.; testimony of Harry W. Hays, U.S. House, Hearings, *Deficiencies in Administration of Federal Insecticide, Fungicide, and Rodenticide Act*, May 7, 1969, 7.

6. U.S. House, Hearings, *Deficiencies in Administration of Federal Insecticide, Fungicide, and Rodenticide Act*, June 24, 1969, 132. The agreement text appears on pp. 70–71.

7. Ibid., 74, 122, 127, 130.

8. Testimony of Harry W. Hays, May 7, 1969, ibid., 23; Juliet Eilperin, "EPA Will Not Have to Consult Wildlife Agencies on Pesticides," *Washington Post*, July 30, 2004, A7.

9. M. J. Sloan and Bernard Lorant to Robert J. Anderson and J. K. Kirk, April 7, 1967, regulatory and control 2, box 16, OA, CCF, 1967–73, ARS, RG 310, NARA.

10. R. J. Anderson to George Irving Jr. and George Mehren, April 24, 1967; Harry W. Hays to Anderson, June 19, 1967, ibid.

11. Harry W. Hays to R. J. Anderson, June 19, 1967; Bernard Lorant and M. J. Sloan to USDA and FDA, June 21, 1967, regulatory and control, box 16; "Report of the Food and Drug Administration's Advisory Committee Appointed to Review Petitions for Tolerances for Residues of Aldrin and Dieldrin on Citrus Fruits, Rice, and Small Grains," committees 1, box 4, ibid.

12. K. R. Fitzsimmons to George Irving Jr., August 22, 1967; Irving to Fitzsimmons, August 25, 1967; L. E. Mitchell to Harry W. Hays, September 15, 1967, regulatory and control, box 16, ibid.

13. Harry W. Hays to F. J. Mulhern, June 12, 1967; E. P. Reagan to Hays, June 21, 1967, regulatory and control 2, box 16, ibid.

14. U.S. House, Hearings, *Deficiencies in Administration of Federal Insecticide, Fungicide, and Rodenticide Act*, May 7, June 24, 1969, 13, 15, 17–19, 23–24, 41, 54; Juliet Eilperin, "Rat-Poison Makers Stall Safety Rules," *Washington Post*, April 15, 2004, A3.

15. U.S. House, Hearings, *Deficiencies in Administration of Federal Insecticide, Fungicide, and Rodenticide Act*, May 7, June 24, 1969, 15–16, 33–35.

16. *Hubbard-Hall Chemical Company* v. *Charles L. Silverman*, 340 F.2d, 402 [1965 U.S. App. LEXIS 6771]. For a case involving a farm worker's chemically induced allergy and employer liability, see *Vanoven* v. *Hardin*, 233 Ark. 301; 344 S.W.2d, 340 [1961 Ark. LEXIS 396].

17. Testimony of Jose Gonzalez, 7–36, in trial transcript, *Jose G. Gonzalez* v. *Virginia-Carolina Chemical Company*, container 169, AC 874, Federal Records Center, Southeast Region, National Archives and Records Center, East Point, Ga.

18. Testimony of Dr. H. E. Farver, 69–86, ibid.

19. Dr. R. M. Gonzalez to Lane and Laney, November 5, 1964, ibid., exhibit P-10.

20. Testimony of William Wannamaker III, 132–33, ibid.; *Jose G. Gonzalez* v. *Virginia-Carolina Chemical Company*, 239 F. Supp., 567 (S.C., 1965) [1965 U.S. Dist. LEXIS 9589].

21. Plaintiff's Trial Brief: Answers to Interrogatories Propounded by Defendant, *Anne Page Griffin* v. *Planters Chemical Corporation*, copy in *Griffin* v. *Planters Chemical Corporation*, civil case 68–170, accession number 72E1507, box 169, location G0781354, National Archives and Records Administration, Southeast Region, East Point, Ga.; *Griffin* v. *Planters Chemical Corporation*, 302 F. Supp., 937 (Ga., 1969) [1969 U.S. Dist. LEXIS 12498].

22. *Griffin* v. *Planters Chemical Corporation,* 302 F. Supp., 937 (Ga., 1969) [1969 U.S. Dist. LEXIS 12498].

23. *International Paper Company* v. *Gilbourn,* 240 S.E.2nd, 722 (Ga., 1977).

24. R. D. Radeleff to H. W. Johnson, September 25, 1962; Radeleff to Johnson, October 10, 1962; Radeleff to Johnson, October 12, 1962; Radeleff, "Report of Consultations and Investigations Concerning Cattle Losses after Spraying or Dipping in Co-Ral," chemicals, box 10, ERD, DC, 1959–65, ARS, RG 310, NARA.

25. R. D. Radeleff to H. W. Johnson, March 15, 1963; Radeleff to Johnson, April 5, 1963, chemicals, box 13, ibid.

26. Nevius A. Stubbs to A. D. Cromartie, July 8, 1965, pesticides, box 4359, GC, 1906–75, SOA, RG 16, NARA.

27. Morton Mintz, "What Directions Do You Follow?" *Washington Post,* July 6, 1969, 33.

28. U.S. House, Report, *Deficiencies in Administration of Federal Insecticide, Fungicide, and Rodenticide Act,* November 13, 1969, 30–33; General Accounting Office, "Need to Resolve Questions of Safety Involving Certain Registered Uses of Lindane Pesticide Pellets, Agricultural Research Service, USDA," February 20, 1969, 3–12; Ned D. Bayley to A. T. Samuelson, November 27, 1968, pesticides, box 5080, GC, 1906–75, SOA, RG 16 NARA; Victor A. Drill, Betram D. Dinman, and Ted A. Loomis to Harry W. Hays, March 17, 1969, box 9; "Registration Policy for Lindane Vaporizers, October 17, 1961," office of the general counsel, box 10, Records of the Environmental Protection Agency, Record Group 412, NARA (hereafter cited as EPA, RG 412, NARA).

29. Mrs. E. H. White to Bureau of Standards, February 16, 1959; Justus C. Ward to Mrs. White, March 30, 1959, office of the general counsel, box 9, landmark litigation case files, 1970–74, EPA, RG 412, NARA.

30. Ben I. Heller to J. T. Herron, October 8, 1959; James C. Munch to Justus C. Ward, March 26, 1960; Munch to Thomas L. Kemp, March 26, 1960, ibid.; General Accounting Office, "Need to Resolve Questions of Safety Involving Certain Registered Uses of Lindane Pesticide Pellets, Agricultural Research Service, USDA," February 20, 1969, 21.

31. J. Philip Loge to Robert A. Rollins, August 14, 1961; D. W. Dean to Justus C. Ward, June 26, 1961, office of the general counsel, box 10, landmark litigation case files, 1970–74, EPA, RG 412, NARA.

32. D. W. Dean to Justus C. Ward, June 26, 1961, ibid.

33. Ibid.

34. Ibid.; Dean to Ward, August 9, 1961; Wayland J. Hayes Jr. to Justus C. Ward, July 24, 1961; Hayes to J. Phillip Loge, August 31, 1961, ibid.

35. Mrs. Louis Kamansky to Thomas H. Kuchel, May 19, 1961; E. D. Burgess to Mrs. Kamansky, June 30, 1961, insects, box 3610; Frank J. Welch to John D. Dingell, March 1, 1962; Orville L. Freeman to Dingell, March 21, 1962; "Organic Consumer Report," January 2, 1962, insecticides, box 3782, GC, 1906–75, SOA, RG 16, NARA; Accidental Poisoning Report, Deborah Kamansky, November 27, 1961, office of the general counsel, box 10, landmark litigation case files, 1970–74, EPA, RG 412, NARA.

36. "Accident Reports Concerning Injuries Resulting from the Inhalation of Vaporized Lindane," n.d., box 9; testimony of Mitchell Zavon, 131, Hearings before Edmund M. Sweeney on Neodane Lindane Vaporizers, June 16, 1971, office of the general counsel,

box 3, landmark litigation case files, 1970–74, EPA, RG 412, NARA; U.S. House, Report, *Deficiencies in Administration of Federal Insecticide, Fungicide, and Rodenticide Act,* 6–7.

37. U.S. House, Hearings, *Deficiencies in Administration of Federal Insecticide, Fungicide, and Rodenticide Act,* May 7, June 24, 1969, 38, 55–56, 58.

38. Lester P. Condon to the Secretary, January 9, 1969, and attachment, "Results of OIG Inquiry into the Pesticide Program"; George W. Irving Jr. to Ned D. Bayley, January 30, 1969, pesticides, box 5080; "United States Department of Agriculture Response to the Recommendations of the Report by the House Committee on Government Operations, 'Deficiencies in Administration of he Federal Insecticide, Fungicide, and Rodenticide Act,'" n.d. (January 1970?), pesticides, box 5270, GC, 1906–75, SOA, RG 16, NARA.

39. George W. Irving Jr. to the Secretary, April 3, 1969; Robert Anderson to Ned D. Bayley, June 16, 1969, pesticides, box 5080, GC, 1906–75, SOA, RG 16, NARA.

40. U.S. House, Hearings, *Deficiencies in Administration of Federal Insecticide, Fungicide, and Rodenticide Act,* May 7, June 24, 1969, 10–11, 13, 59–62, 71.

41. U.S. House, Report, *Deficiencies in Administration of the Federal Insecticide, Fungicide, and Rodenticide Act,* 9–10, 56–59; U.S. House, Hearings, *Deficiencies in Administration of Federal Insecticide, Fungicide, and Rodenticide Act,* June 24, 1969, 113–18; E. F. Behrens to J. Phil Campbell, July 24, 1969, pesticides, box 5080, GC, 1906–75, SOA, RG 16, NARA.

42. Testimony of Mitchell Zavon, U.S. House, Committee on Interstate and Foreign Commerce, *Color Additives: Hearing on H.R. 7624, S. 2197,* 86th Cong., 2nd sess, April 5, 1960, 461–64; U.S. House, Report, *Deficiencies in Administration of the Federal Insecticide, Fungicide, and Rodenticide Act,* 11–12, 62–64.

43. U.S. House, Report, *Deficiencies in Administration of the Federal Insecticide, Fungicide, and Rodenticide Act,* 62–71; Morton Mintz, "U.S. Probing 3 Advisors," *Washington Post,* November 17, 1969, A2.

44. U.S. House, Report, *Deficiencies in Administration of Federal Insecticide, Fungicide, and Rodenticide Act,* 62–70; U.S. House, Hearings, *Deficiencies in Administration of Federal Insecticide, Fungicide, and Rodenticide Act,* June 24, 1969, 78; "United States Department of Agriculture Response to the Recommendations of the Report by the House Committee on Government Operations, 'Deficiencies in Administration of the Federal Insecticide, Fungicide, and Rodenticide Act'"; "FDA Tests Find Shell's Pest Strips are 'Unacceptable,'" *Washington Post,* August 4, 1970, A3; Mintz, "U.S. Probing 3 Advisers," A2.

45. Francis Silver to George L. Mehren, June 23, 1967, pesticides, box 4713, GC, 1906–75, SOA, RG 16, NARA.

46. Harold G. Alford to C. G. Green, September 9, 1969; R. J. Anderson to Charles W. McDougall, September 15, 1969; W. G. Appleby to Alford, September 24, 1969, pesticides, box 5081, ibid.

47. U.S. House, Hearings, *Deficiencies in Administration of Federal Insecticide, Fungicide, and Rodenticide Act,* June 24, 1969, 83–86; George A. Goodling to J. Phil Campbell, July 31, 1969; Harold G. Alford to Robert H. Rowley Jr., September 11, 1969; Campbell to Goodling, October 15, 1969; Glenn M. Anderson to Clifford M. Hardin,

August 25, 1969; Harold G. Alford to Terry Steinhart, August 12, 1969; Ned D. Bayley to Anderson, October 3, 1969, pesticides, box 5081, GC, 1906–75, SOA, RG 16, NARA.

48. L. H. Fountain to Clifford M. Hardin, December 23, 1969, and *Des Moines Register* clipping, James Risser, "Hardin Puzzled; Pesticide Criticism Not Shown Him"; Hardin to Fountain, January 9, 1970, pesticides, box 5270, GC, 1906–75, SOA, RG 16, NARA. On November 21, Ned Bayley had offered to meet with Hardin to discuss the report before he signed the letter acknowledging its receipt. This note was attached to the secretary's reply but did not indicate whether or not the meeting took place.

49. U.S. House, Report, *Deficiencies in Administration of the Federal Insecticide, Fungicide, and Rodenticide Act,* 13–17; George W. Irving Jr. to T. C. Byerly, September 21, 1970, pesticides, box 5269, GC, 1906–75, SOA, RG 16, NARA.

50. Burt Schorr, "Beating the Bans," *Wall Street Journal,* June 26, 1970, 1.

51. Thomas Pynchon, *Gravity's Rainbow* (New York, 1973), 251.

NOTES TO CHAPTER 8

1. Jack A. Warren Jr. to Orville Freeman, February 23, 1967, insects, box 4680; Freeman to Spessard L. Holland, May 25, 1957; George W. Irving Jr. to Freeman, June 20, 1967; "Imported Fire Ant: Sequence of Events," insects, box 4679, GC, 1906–75, SOA, RG 16, NARA. See also R. B. Preacher to Mendel Rivers, August 30, 1967; Freeman to Irving, October 4, 1967, box 4679, ibid. For an excellent account of the mirex campaign, see Buhs, *Fire Ant Wars,* 140–69.

2. W. A. Ruffin to D. R. Shepherd, June 29, 1967, regulatory and control, box 16, OA, CCF, 1967–73, ARS, RG 310, NARA; Richard Carlton to Orville Freeman, January 30, 1968, insects, box 4825, GC, 1906–75, SOA, RG 16, NARA.

3. F. S. Arant to F. J. Mulhern, October 16, 1969, regulatory and control, box 35, OA, CCF, 1967–73, ARS, RG 310, NARA; Buhs, *Fire Ant Wars,* 81–123.

4. Jim Buck Ross to J. Phil Campbell, November 3, 1969; Ross to Clifford M. Hardin, October 27, 1969, insects, box 5026, GC, 1906–75, SOA, RG 16, NARA; Deborah Shapley, "Mirex and the Fire Ant: Decline in Fortunes of 'Perfect' Pesticide,'" *Science* 172 (April 23, 1971): 358–60.

5. Jim Buck Ross to John Orcutt, January 7, 1970, and attached, "Imported Fire Ant," insects, box 5227, GC, 1906–75, SOA, RG 16, NARA. Emphasis in original. See also Ross to Ned Bayley, May 8, 1970, ibid.

6. Nyle C. Brady to William L. Dawson, and attached, "Comments Relative to the Comptroller General's Report: 'Weaknesses and Problem Areas in the Administration of the Imported Fire Ant Eradication Program,'" March 2, 1965, insecticides, box 4309; J. Phil Campbell to Thomas T. Irvin, April 20, 1970, insects, box 5227; C. Edward Carlson to Area Director, plant protection division, ARS, March 24, 1970, pesticides, box 5270, GC, 1906–75, SOA, RG 16, NARA.

7. Louis N. Wise to Ned D. Bayley, May 28, 1970, insects, box 5227, ibid.

8. L. D. Newsom to Ned D. Bayley, June 25, 1970, ibid.

9. W. E. Westlake to S. A. Hall, "Comments on Meeting in Gulfport, December 15, 16, 1959," man and animals affecting, box 1, DC, 1959–65, ARS, RG 310, NARA.

10. George W. Irving Jr. to Ned D. Bayley, June 10, 1970, pesticides, box 5269, GC, 1906–75, SOA, RG 16, NARA.

11. T. C. Byerly to Files, July 10, 1970, pesticides, box 5269; George W. Irving Jr. to J. Phil Campbell, July 28, 1970, insects, box 5227; Byerly to Ned D. Bayley, August 31, 1970, pesticides, box 5269; Byerly to Bayley, December 9, 1970, box 5268, pesticides, GC, 1906–75, SOA, RG 16, NARA; "Suit Seeks to Curb Fire Ant Pesticide in 9 Southern States," *New York Times,* August 6, 1970, 30; Shapley, "Mirex and the Fire Ant," 358. On the ARS's policy toward studies on the carcinogenicity of pesticides, see George W. Irving Jr. to the Secretary, February 4, 1969, box 5080, GC, 1906–75, SOA, RG 16, NARA.

12. Agnes B. Royce to the USDA, October 23, 1970; Mrs. Robert E. C. Weaver to Hale Boggs, November 12, 1970; Mrs. John L. Heard Jr. to Clifford M. Hardin, November 17, 1970; petition from Earth Committee of Spring Hill College, November 23, 1970, mirex responses, box 1, plant protection plan, ARS, RG 310, NARA.

13. Michael Owen Willson to Clifford M. Hardin, September 4, 1970, insects, box 5227; Alice Bailey to J. Phil Campbell, October 10, 1970, pesticides, box 5268; Jerry W. and Sally Nagel to Howard Baker, November 14, 1970; Theron J. Kiskey to Harry F. Byrd, November 23, 1970, insects, box 5227 (emphasis in original); N. P. Raiston to Thomas F. Eagleton, December 31, 1970, pesticides, box 5268, GC, 1906–75, SOA, RG 16, NARA.

14. Jim Buck Ross to W. E. Brock, September 9, 1970, insects, box 5227, ibid. The USDA was being sued in *Environmental Defense Fund, Inc. and Committee for Leaving the Environment of America Natural (CLEAN)* v. *Clifford M. Hardin, et al.*, U.S. District Court for the District of Columbia, Civil No. 2319–70. See Edward M. Shulman to Clifford M. Hardin, August 12, 1970, pesticides, box 5269, GC, 1906–76, SOA, RG 16, NARA.

15. Roy Bailey (Personal and Confidential) to J. Phil Campbell, December 28, 1970, insects, box 5227, GC, 1906–75, SOA, RG 16, NARA.

16. Scott County mimeographed form (emphasis in original); Jessie C. Haley to Richard M. Nixon, January 6, 1971 (emphasis in original); Ronnie Minga to Nixon, January 6, 1971, mirex responses, box 1, plant protection plan, ARS, RG 310, NARA.

17. Elizabeth Brockman to Richard M. Nixon, December 2, 1970; Mrs. Roger W. Brooking to Nixon, December 2, 1970, ibid.

18. Evelyn Angeletti to Richard M. Nixon, December 1, 1970, ibid. See also Hal Bailey to Richard M. Nixon, December 2, 1970; Charles R. Hall Jr. to Nixon, December 2, 1970, mirex responses, box 2, ibid.

19. Martha F. Fielder to the USDA, December 7, 1970 (emphasis in original); Donald G. Legg to the USDA, December 14, 1970; Noral E. Fogle to Richard Nixon, February 22, 1971, mirex responses, box 1, ibid.

20. "U.S. to Spray Controversial Chemical on Fire Ants," *New York Times,* April 12, 1971, 24; Jon Nordheimer, "Environmentalists Fight U.S. Spraying Plan on Fire Ants in South," *New York Times,* December 13, 1970, 78. See also Tom Herman, "Just How Fearsome Is the Fire Ant?" *Wall Street Journal,* February 23, 1971, 1.

21. "Pesticide Mirex Is Approved Conditionally," *Wall Street Journal,* May 5, 1972, 4; Leon W. Lindsay, "Dixie Malady: Ant-Spray Debate," *Christian Science Monitor,* May 5, 1971, 1, 3.

22. Jane E. Brody, "Agriculture Department to Abandon Campaign against the Fire Ant," *New York Times,* April 20, 1975, 46; Buhs, *Fire Ant Wars,* 162–66.

23. For recent articles on environmental issues, see Justin Gillis, "U.S. Will Subsidize Cleanup of Altered Corn," *Washington Post,* March 26, 2003, E1; Jennifer Lee, "Neighbors of Vast Hog Farms Say Foul Air Endangers Their Health," *New York Times,* May 11, 2003, A1; Christopher Drew, Elizabeth Becker, and Sandra Blakeslee, "Despite Warnings, Industry Resisted Safeguards," *New York Times,* December 28, 2003; E5; Manuel Roig-Franzia and Catharine Skipp, "Tainted Water in the Land of Semper Fi," *Washington Post,* January 28, 2004, A3; Rick Weiss, "Engineered DNA Found in Crop Seeds," *Washington Post,* February 24, 2004, A2; Justin Gillis, "Biotech Regulation Falls Short, Report Says," *Washington Post,* April 1, 2004, E3; Juliet Eilperin, "Rat-Poison Makers Stall Safety Rules," *Washington Post,* April 15, 2004, A3; Rick Weiss and Justin Gillis, "Monsanto Beats Farmer in Patent Fight," *Washington Post,* May 22, 2004, A8; Marc Kaufman, "USDA Expands Mad Cow Inquiry," *Washington Post,* July 3, 2004, A2; Rick Weiss, "Nanotechnology Precaution Is Urged, *Washington Post,* July 30, 2004, A2; William Glaberson, "Agent Orange, the Next Generation," *New York Times,* August 8, 2004, 21.

INDEX

Page numbers in italics refer to illustrations and captions.